Sustainable Aviation

T. Hikmet Karakoc • C. Ozgur Colpan
Onder Altuntas • Yasin Sohret
Editors

Sustainable Aviation

 Springer

Editors
T. Hikmet Karakoc
Faculty of Aeronautics and Astronautics
Eskisehir Technical Univeristy
Eskisehir, Turkey

Onder Altuntas
Faculty of Aeronautics and Astronautics
Eskisehir Technical Univeristy
Eskisehir, Turkey

C. Ozgur Colpan
Faculty of Engineering,
Department of Mechanical Engineering
Dokuz Eylul University
Buca, Izmir, Turkey

Yasin Sohret
Airframe & Powerplant Maintenance
Suleyman Demirel University
Isparta, Turkey

ISBN 978-3-030-14197-4   ISBN 978-3-030-14195-0   (eBook)
https://doi.org/10.1007/978-3-030-14195-0

This Springer imprint is published by the registered company Springer Nature Switzerland AG
The registered company address is: Gewerbestrasse 11, 6330 Cham, Switzerland

Preface

Environmental concerns, such as air pollution and global warming, have become a major factor in all industries, also in aviation. Because of using the conventional systems, all problems will dramatically increase. It is necessary to develop more efficient, environmentally friendly, and economical alternative technologies that will be implemented in industries to create a sustainable future and meet the needs of tomorrow's consumers. All parts of aviation industries (aircraft, airports, etc.) have a major impact on pollution and other environmental impacts. Aviation should be considered in detail in terms of new technologies and resources, taking into account the impact of energy, environment, and economy. Here, sustainability, a concept in which all these parameters are evaluated together, is a critical expression. In this book, different sustainability expressions in aviation will be examined.

This book, *Sustainable Aviation: Fundamentals I*, presents and explains the main legs of sustainability in aircraft and airports. This book consists of three parts. In Part I, editors presented the fundamental of sustainability and the contexts of this book. While the aircraft is mentioned in Part II, airport is told in Part III, according to sustainable approaches and methods. Four subheadings (Energy Management, Risk Methodology to Assess and Control Aircraft Noise Impact in Vicinity of the Airports, Biodiversity Management, and A Holistic View of Sustainable Aviation) is stated in Part II. Two subheadings (A Short Brief on the Aircraft History and Anatomy, and Technology Review of Sustainable Aircraft Design) are stated in Part III.

Eskisehir, Turkey   T. Hikmet Karakoc
Buca, Izmir, Turkey   C. Ozgur Colpan
Eskisehir, Turkey   Onder Altuntas
Isparta, Turkey   Yasin Sohret

Contents

Part I
Introduction

Chapter 1
Fundamentals of Sustainability

Onder Altuntas, Yasin Sohret, and T. Hikmet Karakoc

Technology has increased due to the fast growth of the population, because of human demand. Man is in a kind of examination with his environment and world. At this point, the concept of sustainability needs to be understood. It is desirable to understand sustainability and its effects in aviation with this book. In this chapter, basic concepts about sustainability are described.

1.1 What Is the Sustainability?

There are several definitions for sustainability. The main idea of them is defined as [1]:

> The ability of the system to maintain the current state when exposed to external influences. If the current state is not saved, then this state is called unstable.

All areas (not only engineering) have their own definition. For example, while sustainability is defined as the property of technical systems to maintain the values of design and operating parameters within specified limits in technology, it also refers to the long-run balance between resource exploitation and human development in macroeconomic. Moreover, it can define all areas of life.

While the word of sustainability was firstly used in 1987 [1], the words of "sustainability" and "aviation" are firstly used in 1998 [2]. The definition of sustainable

O. Altuntas (✉) · T. H. Karakoc
Eskisehir Technical University, Faculty of Aeronautics and Astronautics, Eskisehir, Turkey
e-mail: oaltuntas@eskisehir.edu.tr; hkarakoc@eskisehir.edu.tr

Y. Sohret
Suleyman Demirel University, Isparta, Turkey
e-mail: yasinsohret@sdu.edu.tr

© Springer Nature Switzerland AG 2019
T. H. Karakoc et al. (eds.), *Sustainable Aviation*,
https://doi.org/10.1007/978-3-030-14195-0_1

aviation is clearly explained in the statement of sustainable transport as "basic mobility to all citizens without damaging nature and the environment" [2].

Aviation is one of these areas. Sustainability characterizes the ability of an aircraft without the intervention of a pilot to maintain a given flight mode in aviation. Besides, there are several fields in aviation, such as aircraft, airport, management, etc. So, like different sectors, all subfields have their own sustainability definition in aviation sector.

1.2 Evaluating of Sustainability

Sustainable development is defined as sustainable development strategy in England in 1999 [2, 3]. This strategy, the integrated economy, can be explained as environmental and social policies providing the best quality for everyone in the present and next generations.

At this point, four titles can be listed:

- Sustainability of high and fixed levels of economic development and work
- Effective protection of the environment
- Use of natural resources
- Social progress needed by everyone

A simple formulation has been developed by HMG. It is "a better life for now and the next generation." Another definition is that "the best environmental performance is better for the next generation and for the future generations."

Development is meeting today's needs without compromising the ability to meet the needs of future generations. To develop in near future, sustainability is a very important key. There are three pillars for the sustainability, which are economic, social, and environmental issues [4].

1.2.1 Economic Sustainability

While economic development is to give people what they want and endangers the quality of life [4, 5], economic sustainability means that we must use, safeguard, and sustain resources (human and material) to create long-term sustainable values by optimal use, recovery, and recycling. In other words, we must conserve finite natural resources today so that future generations too can cater to their needs.

1.2.2 Social Sustainability

The most important is the awareness and protection of people's health against pollution and other harmful activities of companies and other organizations [4]. Since discussions on sustainable development are often focused on the ecological or economic aspects of sustainability, social sustainability is an aspect of sustainability that is often overlooked. Three dimensions of sustainability should be addressed to achieve the most sustainable outcome possible.

1.2.3 Environmental Sustainability

Humans should protect the environment, for their future, whether it's about recycling or reducing our energy consumption by switching electronic devices instead of stationary or by making short trips instead of taking the bus. Here sustainability is an important step. Environmental sustainability is the rate of extraction of renewable resources, generating pollution, and nonrenewable depletion of resources that can be sustained indefinitely. If they cannot be pursued indefinitely, they are not durable [5, 6].

Apart from these three basic development definitions, described above, there are many different developmental values. All of them are directly related with these three pillars, economic, social, and environmental. For instance, definition of "sustainable aviation" should be solved with economic, social, and environmental sustainabilities.

References

1. Brundtland GH (1987) Report of the World Commision on Environment and Development "Out CommonFuture", pp. 11–17, https://sswm.info/sites/default/files/reference_attachments/UN%20WCED%201987%20Brundtland%20Report.pdf
2. Sledsens T (1998) Sustainable aviation: the need for a European environmental aviation charge, T&E 98(1). European Federation for Transport and Environment, Brussels
3. IPCC (1999) IPCC special report. Aviation and the global atmosphere: summary for policymakers. https://www.ipcc.ch/report/aviation-and-the-global-atmosphere-2/
4. Adams WM The future of sustainability: re-thinking environment and development in the twenty-first century. Revised: 22 May 2006, http://cmsdata.iucn.org/downloads/iucn_future_of_sustanability.pdf. Accessed 10 Sep 2018
5. Paschek F (2015) Urban sustainability in theory and practice-circles of sustainability. Town Plan Rev 86(6):745
6. EPA. http://www.epa.gov/sustainability/basicinfo.htm. Accessed 10 Sep 2018

Part II
Airports

This part consists of four chapters. They are "Energy Management," "Risk Methodology to Assess and Control Aircraft Noise Impact in Vicinity of the Airports," "Biodiversity Management," and "A Holistic View of Sustainable Aviation."

While chapter authors describe the management of the energy in airports in the first chapter, the management of the biodiversity is investigated in the third chapter. Airport noise impact and a holistic view of sustainable aviation are described in second and fourth chapters, respectively.

Chapter 2
Energy Management at the Airports

M. Kadri Akyuz, Onder Altuntas, M. Ziya Sogut, and T. Hikmet Karakoc

The number of airports has recently increased due to the fast growth of the aviation industry and the increment in the number of new destinations added to the flight network. The dynamic structure, 24-hour servicing, performing flight operations safely, and the provision of comfortable conditions for passengers in terminal buildings, has led to an increase in the energy consumption of airports. Energy management has become an important topic for the continuous improvement of energy performance in airports as they are sites with high and intense energy consumption. In this chapter, an energy management model is proposed for airports considering sustainable development principles. The proposed energy audit models are applicable to an international airport terminal building.

2.1 Introduction

The rapid growth of the aviation industry (by an annual average of 5%) and the addition of new routes to the flight network in recent years have led to an increase in the number of airports. The fact that the aviation industry makes a contribution of approximately 2.2 trillion dollars to the world economy with more than 3.3 billion passengers per year is also an indication of to what extent it is great [1]. The main

M. K. Akyuz
Dicle University, School of Civil Aviation, Department of Airframe and Powerplant
Maintenance, Diyarbakır, Turkey
e-mail: mkadri.akyuz@dicle.edu.tr

O. Altuntas (✉) · T. H. Karakoc
Eskisehir Technical University, Faculty of Aeronautics and Astronautics, Eskisehir, Turkey
e-mail: oaltuntas@eskisehir.edu.tr; hkarakoc@eskisehir.edu.tr

M. Z. Sogut
Piri Reis University, Istanbul, Turkey

© Springer Nature Switzerland AG 2019
T. H. Karakoc et al. (eds.), *Sustainable Aviation*,
https://doi.org/10.1007/978-3-030-14195-0_2

reasons for this increase are the increase in passengers' demand for air transport and the addition of different points (city/country) to the flight network. The inclusion of different flight points in this network has led to an increase in the number of airports and relevant facilities, as well.

Airports have a dynamic and complex structure that contains many service sectors within themselves. The main focal point especially in ground services is terminal buildings. However, dynamic processes and services, the systems or elements for the safe realization of flight operations, and ensuring comfort conditions of passengers in terminal buildings at the airports are basically considered as the components that are directly connected to energy resources. Nowadays, the fact that airports achieve a sustainable infrastructure in environmental impact processes is primarily possible by the development of effective and efficient energy management systems. Indeed, as it is seen in sectoral examples, energy management has become an important issue for the continuous improvement of energy performance at airports which are high and intensive energy consumption areas.

In this section, energy management issue at airports (under sustainability) was primarily discussed by taking into account the sustainable principles. An attempt to express a sample model for airports was made with reference to ISO 50001 EnMS (energy management system) which is the basis of energy management. Preliminary and detailed energy audits, which will form a basis for the actions to be created in order to ensure sustainable effect, were performed through reference international airports. To address the energy issue at sustainable airports in a holistic approach and to take into account the sustainable development principles of all energy-based effects were discussed in detail in this section.

The increase in the number and capacity of the airports primarily leads to an increase in demand for energy for the airports. When it is considered around the world, it is seen that a large part of this energy requirement at airports is provided from fossil fuel sources, like coal, natural gas, and oil. It is known that fossil fuel-based energy consumption and this ever-increasing demand lead to economic, social, and environmental problems, as well. Other energy resources, including hydroelectric, are considered to be renewable and therefore sustainable in the long term. The energy generated from renewable resources does not mean that it has zero emission and is environmentally harmless. A certain amount of damage is caused to the environment in the processes of obtaining the raw materials of renewable energy systems from nature and producing, installing, maintaining, and recycling them [2, 3].

The fact that fossil fuel resources are decreasing rapidly, are not sustainable, and are the main reason of global warming problem requires the use of energy more rationally and efficiently [4]. In recent years, the legal obligations arising from global warming and climate change have led to the need for using energy more efficiently in all sectors. The share of transport sector in carbon emission, which is a measure for climate change and global warming threat, is 26.4%. The International Civil Aviation Organization (ICAO) estimates that aviation-induced CO_2 emissions have a share of 2% and that this will increase by approximately 3% to 4% in every year [5, 6]. Airports are responsible for 5% of aviation-induced CO_2 emissions [7].

Energy management is the most effective way to prevent energy wasting. While there is a potential for energy efficiency in almost every sector around the world, it is possible to improve the energy potential by 40% even in the countries that use energy most efficiently, with current technology and efficient energy management [8]. Energy management should be addressed holistically to decrease energy intensity and energy-induced environmental impacts without compromising comfort and quality of service at airports which are high and intensive energy consumption areas.

The segmentation of the airport to be made most generally will be important for the establishment of the energy management approach. Airports can be examined in two sections: air side and land side. The section called land side consists of terminal building, cargo terminal, and parking area. The air side section (apron, runway, control tower, maintenance facilities, etc.) includes all areas and structures related to aircraft [9].

2.2 Sustainable Energy Management

Despite all these structures, energy continues to be an important problem of sustainability in social development. For this reason, it has become a necessity to control energy consumption, to make sustainable energy policies, and to address them in a holistic manner. In this context, it is primarily seen that the components of sustainable energy management system are improvable resources for strategies. Sustainable energy management consists of power management, technology management, risk management, policies, sustainability, awareness management, environmental management, and resource management [10].

Energy, as an economic and environmental actor in sustainable development strategies and practices, is an issue that needs to be taken into consideration both the cost potential created by it and its environmental impact aspect. Energy is a basic input for all components or sectors in society. Only the building sector has approximately 40% of the total energy consumption potential. Therefore, the development of holistic approaches has gained importance for sustainable assessments in decision processes. This concept has made it valuable to develop energy management systems and to increase the effective and efficient use of energy. The effective and efficient use of energy in energy management and the development of more cost-effective solutions have made holistic approaches as a necessity for sectoral or subsectors as well as national strategies. Airports, which have a significant role in energy consumption, should be considered as the areas where energy is not used correctly and effectively. Energy management should be considered as a requirement in all local or integrated elements in this sector where almost all energy consumption tools can be used.

Along with the general management approaches of nations, ISO 50001 Energy Management System (EnMS), which is valued internationally, is an effective program which has been used as an improved management system program from industry to buildings. In this context, ISO 50001 aims to create an energy-efficient management infrastructure based on continuous improvement. The effective energy system approach, as well as the holistic management of airports, will have a direct or indirect sustainable effect. Thus, ISO 50001 EnMS is a management system for decreasing energy costs for businesses. In this context, all components of the system were separately examined below.

2.2.1 ISO 50001 Energy Management System

Energy management is a system and process developed by organizations to improve energy performance, including the efficiency, consumption, and use of energy [11]. ISO 50001 EnMS forms the basis of energy management. ISO 50001 has been based on the management system model implemented worldwide. ISO 50001 management strategies increase the energy efficiency of public and private sector organizations, decrease the costs, and improve energy performance. ISO 50001 standards have been designed to enable organizations to integrate their energy performance into management practices within a known framework. ISO 50001 EnMS forms a framework related to the subject and follows the PDCA (plan-do-check-act) processes as in other management systems [12].

The PDCA is a common cycle used in all management systems and consists of four stages. These are [13]:

Plan: Planning the system, process, and resource allocation to achieve the objective.
Do: To carry out the necessary actions in accordance with the plan.
Check: Measurement, monitoring, and verification of results in accordance with the predefined criteria.
Act: Analyzing the results and important changes necessary to improve the process.

The EnMS model is presented in Fig. 2.1 [14]. The benefits of ISO 50001 EnMS are listed below [15, 16]:

- To guide the management of energy resources.
- To ensure transparency and to facilitate communication in the management of energy resources
- To promote best energy management practices and to strengthen energy management behaviors
- To help the evaluation and prioritization of energy-efficient new technological practices
- To create a framework to promote energy efficiency throughout the supply chain
- To enable energy management practices for greenhouse gas emissions reduction projects
- To allow integration with other organizational management systems such as environment, health, and safety

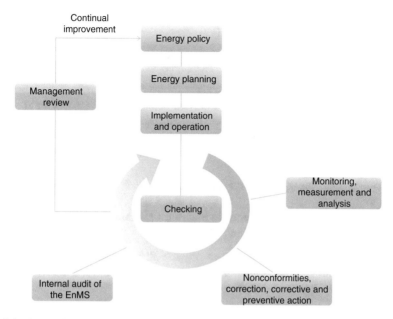

Fig. 2.1 ISO 50001 energy management system model. (Adapted from: ref. [14])

ISO 50001 has been designed to be implemented regardless of the size and area of activity of any organization, whether it is private or public sector, and its geographical location, like all other ISO management system standards. Businesses can decrease their energy consumptions and improve their energy performance regardless of the size and area of activity. It is possible to take it further by top management commitment on energy efficiency, awareness of all employees, and the use of the right resources in their activities [11].

2.2.1.1 Top Management Commitment

An energy management system cannot succeed without the commitment of top management in this respect. First, the top management should establish an energy policy and identify the resources for the development, implementation, maintenance, and continuous improvement of EnMS. Top management should define the scope and boundary of EnMS and communicate the importance of energy management to all employees and stakeholders. In addition, top management should assign an energy representative and enable the establishment of an energy management team. The management representative and the energy team will be responsible for the energy management activities. Without the commitment and support of the management, EnMS is not implemented successfully, and there is no improvement in energy performance. The commitment of top management is an essential fact in

EnMS as in all management systems. The role and responsibilities of the top management are summarized below [17]:

- Implementation, development, and maintenance of EnMS
- The review and approval of the necessary documents written and published by the EnMS team
- Appointing a management representative and energy management team
- Providing adequate resources for the implementation of EnMS
- Conducting management review of EnMS
- Announcing the importance of energy saving and to encourage all employees in this regard
- Ensuring that energy performance indicators (EnPIs) are appropriate to the airport
- Considering energy performance in long-term planning
- Considering energy performance while deciding

Top management should commit to provide all necessary resources, including the following items, for the implementation of EMS. These resources include [16]:

- Human resources
- Expertise skills
- Technology
- Financial resources

2.2.1.2 Management Representative

The management representative is the person who has been duly authorized by the top management on all energy issues and is accountable to the top management. The appointment of management representative means the representation of EnMS by the top management, which is also a part of the commitment of top management for the improvement of energy performance. The management representative should have knowledge, experience, and management skills on energy and sustainability.

2.2.1.3 Energy Manager

The energy manager is defined as "the person responsible for the fulfillment of energy management activities" [18]. In a sustainable energy management system at airports, the duties of the energy manager can be listed as follows:

- Determining the energy-saving potential.
- Determine the opportunities for energy improvement.
- Determine energy-inefficient equipment with energy studies and to create action plans related to them.
- Identifying the actions for energy efficiency in the short, medium, and long term.
- Monitoring the energy performance continuously.

- To ensure that all activities performed with respect to energy are recorded.
- To include the energy team in all studies related to EnMS and take into account the recommendations of the energy management team regarding all of the items written above.
- Increasing the motivation of the energy management team.
- Preparing reports on financial investments for energy-efficient technology.

2.2.1.4 Energy Team

The energy team consists of individuals who are responsible for the effective implementation of EnMS's activities and for the improvement of energy performance. The role of the energy management team is to monitor and improve energy performance and communicate this to the top management. At airports, the energy management team should be formed by taking into account minimum requirements. The number of team member to serve in the energy management team should be determined by considering size and energy structure of the airport. The fact that the members in the energy management team to be established for continuous improvement of energy performance have necessary knowledge and experience according to their responsibilities within the EnMS, are trained on the relevant subject, and work voluntarily within the system is important for an improved energy performance. The energy management team should include the following corporate areas.

- Engineering
- Airport maintenance/operation
- Procurement
- Training
- Finance
- Human resources
- The law department to monitor compliance with legislative regulations

ISO 50001 does not provide detailed information about the energy management team and number. The energy management team should be formed by taking into account the organizational structures and activities of the airport. When it is considered in terms of airports, the establishment of minimum engineering, airport maintenance/operation, procurement, training, and law departments is important for the following aspects:

Engineering: It is important in terms of performing all engineering (economic, environmental, etc.) analyses on energy at the airport, planning, conducting, and interpreting energy studies, continuous monitoring of energy performance, creating the maintenance plans for energy-consuming equipment, and carrying out research and development activities.

Airport maintenance/operation: It is important in terms of immediate response to failures and irregularities that may occur in energy-consuming systems and performing the planned corrective and preventive actions.

Procurement: It is important in terms of the establishment of pro-forms related to the monitoring and purchasing of energy-efficient equipment available in the market and monitoring the purchase process.

Training: To save on energy without making any financial investments at airports is only possible when all employees have the necessary awareness in this respect. This awareness should be created, and all employees should have knowledge about the benefits of an improved energy performance. All employees should receive necessary training on the required subjects for the formation of a corporate culture related to energy efficiency. The establishment of training programs for all employees working at the airport, providing necessary trainings, and recording them are important for the energy team to follow the required ISO 50001 training.

Law: It is important in terms of monitoring the legal regulations related to energy and environment at the airports.

2.2.1.5 Energy Policy

Energy policy should commit the improvement of energy performance of the organization. The energy policy should be defined by the top management, and this policy should include the following commitments [16]:

- Continuous improvement of energy performance
- The use of all the necessary information and resources to achieve the objectives and targets
- The commitments related to the compliance with the legal and other requirements regarding the use, consumption, and efficiency of energy

The top management approves energy policy. Energy policy is important, as it is a written document of the top management's activities related to energy efficiency and its support and commitment to the EnMS.

2.2.1.6 An Exemplary Energy Policy

We commit to continuously improve the energy performance based on the principles of sustainable development during the activities and services we have performed. While implementing this commitment:

- Using the energy resources in the most efficient way without compromising service quality and comfort.
- Consider the principles of sustainable development in all energy-related activities.
- Complying with national and international legislations, regulations, and laws.
- Continuously monitor the energy consumption, provide all the necessary information and resources to achieve the objectives and targets.
- Provide integration of energy management system with other management systems.

- To make energy-efficient and environmentally friendly choices in energy-related investments.
- Continuously monitor and evaluate our aims and objectives and to revise.
- Providing the necessary resources for the establishment of corporate culture related to the EnMS.
- Evaluating all employees' contributions, initiatives, and ideas related to the EnMS.
- Reducing fossil fuel consumption by 20% within the next 3 years.
- Providing all employees with the necessary training to ensure that the EMS is effectively implemented, continuously improved, and fully understood.

2.2.2 Energy Planning

The energy planning process should be developed considering the national and international legal and other requirements to which the airport is subjected. All data related to energy planning should be collected and analyzed until useful results are obtained [17].

The purpose of energy planning is to obtain planning outcomes. These outcomes are [16]:

- Energy baseline
- Energy performance indicators (EnPI)
- Objectives
- Targets
- Action plans

Along with the collection of past and current data on energy use and consumption, the energy resources used at the airport and in which equipment and processes these resources are used should be determined. In particular, the determination of high and intensive energy consumption points is important in terms of efficiency potential and improvement opportunities at these points.

By analyzing energy data, energy maps should first be created, and significant energy users (SEUs) should be determined. Where, for what purpose, and how much energy is consumed at the airports should be expressed with a single unit (kWh, cal, etc.). While determining significant energy users, the users with the highest share in total energy consumption should be selected.

Energy performance indicators (EnPI) are determined as a result of the statistical analyses performed using the past energy consumption data and the other variables at the airport (number of passengers, number of flights, HDD, CDD, load, etc.). With the equation to be obtained as a result of the regression analyses to be performed between total energy consumption and the variables, both EnPIs can be determined and future energy consumption estimates can be made. The most important issue to be considered in statistical analysis is to determine the variables that affect energy consumption with high reliability. The most important indicators of analysis outputs are corrected R^2 value and p value. The corrected R^2 value shows the degree of the relationship between energy consumption and variables; the closer this value is to 1, the stronger the relationship between energy consumption and variables is. For instance, the fact that the corrected R2 value is found to be 0.9 means that the selected variables explain the energy consumption by 90%. Furthermore, p value in regression analyses is important in terms of whether the relationship between the relevant variables in the selected model and energy consumption is significant. This value is usually required to be less than 0.05.

Energy baseline appropriate to the structure of the airport should be established with the past and current energy data. Energy baseline can be established for each resource or for each variable that affects energy consumption. Energy baselines are very useful tools in terms of monitoring the energy performance and comparing it with previous periods after the action plans established in accordance with the determined objectives and targets are implemented. It is necessary to monitor whether the saving determined by engineering calculations made in the action plans implemented has been actually achieved. If the theoretically calculated savings ratio has not been realized, its reasons should be investigated, and the relevant measures should be taken.

The objectives and targets determined should be in accordance with the energy policy of the airport, and these objectives should be [19]:

- Specific
- Measurable
- Achievable
- Realistic
- Timely

Targets determined in accordance with the objectives should be applicable to the airport. The objective of energy management is the continuous improvement of energy performance. For instance, to decrease energy consumption at the airport by 20% until the end of the year 2018 and based on the year 2016 is an objective. To decrease natural gas consumption by 20% and reduce electricity consumption by 20% is a target to accomplish this objective. Action plans are created and implemented by taking into account the improvement opportunities determined based on preliminary and detailed energy audits in order to achieve these objectives.

The order of priority should be taken into account while implementing the action plans in a sustainable EnMS. It is recommended to follow the following order while prioritizing the action plans:

- Plans that do not require any investment
- National and international legal requirements
- Easy and short-time applicability
- High energy performance with low cost
- High environmental performance
- Economic and environmental payback periods
- Plans related to significant energy users
- Plans related to renewable energy, if available

2.2.3 *Do*

The action plans and other outputs determined in the energy planning stage are used in the implementation process. The stages of the implementation process are presented in Fig. 2.2.

2.2.3.1 Awareness, Training, and Competence

For an improved energy performance, all employees should be aware of the EnMS, energy policy, and their duties and responsibilities within the EnMS. It is necessary to raise the awareness of all employees at the airport about the presence of EnMS and the benefits of an improved energy performance to ensure this awareness. It is possible to raise the awareness of airport employees by the methods such as brochures, training documents, and training videos prepared on the subject. Furthermore, awareness can be guaranteed by adding initial trainings on the EMS to the compulsory trainings that airport employees should receive. The fact that all employees who will affect energy performance are experienced, well informed, and competent on the relevant subjects is the most important fact for the implementation of the EnMS and the continuous improvement of energy performance. The training needs of the energy management unit and the employees affecting energy performance should be determined, and they should be provided with the relevant trainings. In particular, the fact that the employees on significant energy users are competent in this field is one of the essential cases for the continuous improvement of energy performance.

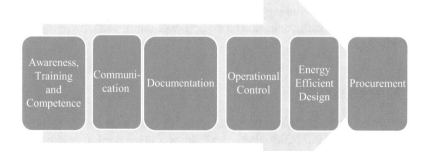

Fig. 2.2 Do. (Adapted from Ref. [16])

2.2.3.2 Communication

Communication is one of the most important processes of the implementation stage. It is necessary to determine which methods will be used for internal communication related to EnMS. It is possible for all employees to be aware of the studies carried out on the EnMS by internal communication. With respect to internal communication, the structure and size of the airport should be considered. The following methods can be used for internal communication at the airports [17]. Furthermore, different methods can be selected and developed in order to provide internal communication.

- E-mail
- Brochure
- Journal
- Internet
- Meeting
- Seminar
- Video
- Advertising boards, etc.

In particular, to encourage all employees to deliver their opinions about the EnMS can make a significant contribution to the improvement of energy performance. The methods to communicate the opinions and suggestions of all employees about the EnMS should be determined by considering the structure and size of the airport. In particular, the creation and encouragement of a standard reporting form contributes to the establishment of this culture at the airports.

The requirements and methods for the communication of the EnMS to the airport energy suppliers, stakeholders, and various organizations operating at the airport should be determined. The fact that airport energy policy and energy performance are open to external communication is important in terms of corporate image. If it has been decided to share the energy performance and the realized action plans with the customers, stakeholders, and interested parties, annual reports on energy performance, energy-saving projects, and energy costs can be made available to external communication on the corporate internet address or in the form of activity reports.

2.2.3.3 Documentation

Energy management activities should be recorded and preserved for an effective EMS. According to ISO 50001 EMS, the following items need to be documented [16, 17]:

- Scope and boundary
- Energy policy
- Energy planning process

- Criteria and methods for reviewing energy
- Methods for determining and updating the EnPIs
- Energy objectives and targets
- Action plans
- Explanation of basic elements of EnMS and the interactions of them
- Document control
- Energy purchasing specifications
- Plans and programs for internal audits
- Other documents determined to be necessary by the organization

According to ISO 50001, the following records need to be preserved [16]:

- Energy review
- Opportunities for improving energy performance
- Energy baselines
- Records of competence training, and experience
- Trainings provided to meet competence requirements and other activities
- External communication decision
- Results of design activities
- Monitoring and measurement results of key activities
- Calibration records of the equipment used in the monitoring and measurement of key activities
- Significant deviations occurring in the energy performance
- Records related to the evaluation of compliance with legal and other requirements
- Records of internal audit results
- Records of corrective and preventive actions
- Reviewing of the management

2.2.3.4   Operational Control

The preparation and implementation of operating instructions by taking into account the significant energy users and the design parameters of the systems and equipment significantly affect the energy performance. It is necessary to carry out the regular maintenance of them, to continuously monitor the energy consumption, and to carry out preventive maintenance if it is necessary. Maintenance procedures and periods related to significant energy users, systems, and equipment should be established, and maintenance activities should be carried out at the specified intervals.

2.2.3.5   Energy-Efficient Design

Energy-efficient design should be selected by taking into account the airport energy policy and energy performance during the construction of new facilities or renewal of existing facilities. In addition to this, environmental criteria should also be considered in energy-efficient designs in sustainable energy management. The stage of the construction of new facilities or renewal of existing facilities should be evaluated by taking into account the environmental, economic, and social aspects of sustainability. All action plans should be implemented in accordance with the principles of sustainable development. Action plans should be developed by taking into account the energy and environmental performance during the design process. The best option can be determined by performing the economic and environmental analysis of this process using the Life Cycle Cost (LCC) and LCA.

2.2.3.6 Procurement of Energy Services, Products, Equipment, and Energy

In the purchase of energy, services, tools, and equipment that will affect significant energy users or energy performance, suppliers should be notified that all these purchases are evaluated on the basis of energy performance. Furthermore, the price, performance, and efficiency criteria for these purchases should be investigated by the procurement department and should be indicated in the purchase specifications. All purchases related to energy should be considered as an opportunity to improve energy performance and should be evaluated and carried out on this basis. In the purchase of energy-efficient equipment, the energy performance and environmental impact of the equipment during their life cycle should be evaluated, and the purchase specifications should be created by taking into account the life cycle of the equipment. It is also important to evaluate the opportunities that will decrease the costs and reduce environmental impact while supplying energy. The criteria that need to be considered in energy supply are summarized below [20].

- Energy quality and availability
- Capacity and cost
- Environmental impact
- Renewability
- Other parameters deemed appropriate by the organization

2.2.4 Check

In energy management system, the check process consists of monitoring, measurement and analysis, evaluation of compliance with legal requirements, internal audit of the EnMS, determination of nonconformities, corrective and preventive actions, and the control of records.

2.2.4.1 Monitoring, Measurement, and Analysis

The fact that the key characteristics that determine the energy performance are regularly monitored, measured, analyzed, and recorded is a requirement of the standard. These characteristics should at least include [21]:

- Significant energy uses and energy review outputs
- The variables related to significant energy users
- Energy performance indicators
- Effectiveness of action plans for achieving objectives and targets
- Actual and expected energy consumption

Continuous monitoring of energy consumption and regular control of EnPIs and SEUs are important tools for determining the effectiveness of the EnMS. The effectiveness of all activities carried out in the previous stages is checked in this stage. To be sure of the accuracy of devices and meters that measure energy consumption is an important parameter for monitoring energy consumption and performance. The calibration requirement and range of measuring devices and meters should be determined, and they should be calibrated if necessary. Furthermore, the fact that the energy efficiency measuring devices used for performing detailed energy audits are calibrated is important in terms of measurement accuracy. These devices should have calibration protocols, and they should be calibrated at the intervals specified by the producing company. All documents related to calibrations should be recorded and preserved.

The effectiveness of the action plans determined to achieve objectives and targets in the energy planning process should be checked in this stage. It is necessary to be sure that the energy savings targeted while creating action plans are realized after the action plans are implemented. If energy savings calculated by engineering analyses performed while creating action plans are obtained, the determined objective has been achieved. However, if there is no improvement in energy performance after the action plan has been implemented, the reasons for it should be investigated, and necessary measures should be taken.

The expected energy consumption (standard energy consumption) obtained by regression analysis is interpreted by comparing with the actual energy consumption. The ratio and amount energy saving can be determined by the results obtained. If the value obtained by subtracting the actual energy consumption from the expected energy consumption is negative, it means that energy has been saved, and the value between them refers to the amount saved. If there are significant deviations and undesired results between the expected energy consumption and the actual energy consumption, the root causes of them should be investigated, and the necessary measures should be taken.

2.2.4.2   Evaluation of Compliance with Legal Requirements

The energy team should periodically evaluate the legal and other requirements that need to be complied at the airport. The cause of any deviation should be analyzed, and corrective measures should be taken. The evaluations are [20]:

- Forwarded to the management representative
- Used as an input in management review
- Recorded and preserved

2.2.4.3   Internal Audit of the EnMS

Internal audit is a process used to objectively evaluate the current situation in order to determine the compliance of the EnMS with the objectives and targets set and to ensure that it is effectively implemented and sustained. The internal audit process should be systematic, independent, and objective [16]. The internal auditor can be from the energy management team or by competent external resources. The most important issue to consider here is that the internal auditor is competent in this respect. The goals of internal audit are sustaining the EnMS, raising awareness, and determining the difference between planning and implementation. Internal audit should be performed at least in every 12 months by taking into account the requirements and the energy structure. An internal audit plan and calendar should be established and systematically implemented before performing internal audit. If audits that were performed previously are available, their outcomes should be taken into account. The reasons for the nonconformities determined during the internal audit process should be investigated, and the necessary corrective actions should be taken. The internal audit results are recorded and reported to the top management.

2.2.4.4   Nonconformities, Corrective, and Preventive Actions

In the energy management system, it is a requirement to determine the existing and potential nonconformities and to implement the corrective and preventive actions related to them. An existing nonconformity is a situation where any requirement has not been met. Potential nonconformity is a situation where there will be nonconformity in the future if a measure is not taken. After the nonconformity is determined, the type of nonconformity (minor or major) and its effect on energy performance should be determined [22].

Minor nonconformity: It can be defined as the deviation in one of the requirements of the standard that does not interfere with the implementation of the EnMS.
Major nonconformity: It can be defined that the EnMS has no ability to meet the requirements of the standard or as a complete collapse of the system.
After the nonconformity is determined, the root cause of the nonconformity is investigated, and its cause is determined. An action plan (corrective/preventive action) should be prepared, implemented, and recorded to eliminate the

nonconformity. After the action plan is implemented, the effectiveness of the action plan should be evaluated, and it should be verified that the nonconformity has disappeared.

Corrective action: It can be defined as the elimination of the cause of an existing nonconformity in the EnMS.

Preventive action: It can be defined as the elimination of the cause of a potential nonconformity in the EnMS.

2.2.4.5 Control of Records

One of the requirements of ISO 50001 EMS is the creation and storage of all records specified in Sect. 2.2.3.3 of this chapter. One of the most important indicators of the implementation of the EnMS is the records required by the standard.

2.2.5 Reviewing of the Management

The actions to be taken in the process of reviewing the EnMS are, respectively, presented in Fig. 2.3. The management review process is determined by the top management by taking into account the size and activities of the airport. The primary purpose of the management review process is to evaluate the effectiveness of the EnMS and to make changes if necessary.

For the continuous improvement of energy performance, it is a requirement to review the EnMS at least once a year to secure its compliance, competence, and effectiveness. The review of the EnMS is performed by the top management. The management review entries mentioned in Fig. 2.3 are evaluated. Management reviews are an important part of the "check" and "act" stages for the continuous improvement of energy performance. The necessary changes related to EnMS are made if it is necessary at the end of the process. These changes are also called as the management review outcomes. These are the [17]:

- Energy performance
- Energy policy
- EnPIs
- Objectives and targets
- Changes in the allocation of resources

2.3 Energy Management at the Airports and a Sample Application

Airports consume almost as much energy as a small city [22]. The airside energy consumers basically consist of radio navigation systems, control tower, apron lighting, and runway lights. In addition to this, aircraft maintenance hangars,

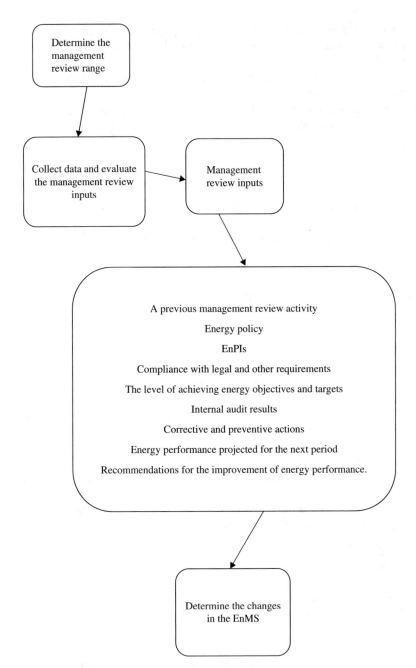

Fig. 2.3 Management review process

Fig. 2.4 Distribution of an airport's energy consumption

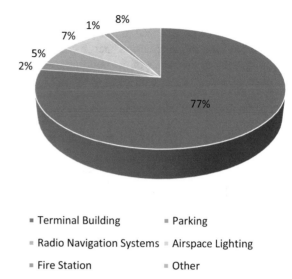

- Terminal Building
- Radio Navigation Systems
- Fire Station
- Parking
- Airspace Lighting
- Other

Fig. 2.5 Typical energy consumption at a terminal building

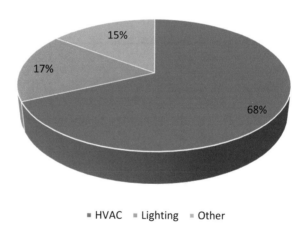

- HVAC
- Lighting
- Other

meteorological stations, and fire stations are important energy users at the airports. The factors that mostly affect the energy consumption on the airside are the length of the runway to be illuminated, the length of the taxiway, and the size of the apron area [9]. The capacity and size of the aircraft maintenance hangars on the airside can significantly increase energy consumption.

The landside basic energy consumer is the terminal building as it is seen in Fig. 2.4. The terminal building is the key point for passengers and cargo to reach air transport. HVAC, lighting, and information and communication technologies systems consume too much energy in terminal buildings [23, 24].

Terminal buildings are generally the structures with very large areas and volumes, and to provide comfort conditions in these structures leads to higher energy consumptions. The most energy-intensive system at airports is the HVAC system as shown in Fig. 2.5. Terminal buildings generally have areas and volumes that are too large, and comfort conditions cause higher energy consumption. Thus, HVAC systems must be emphasized in energy efficiency studies of airport terminal buildings. Regarding the quantity of energy consumed for air-conditioning in terminal buildings, the performance of the building envelope also has a very important role, as does the efficiency of the HVAC system [25, 26].

2.3.1 Factors Affecting Energy Consumption at the Airports

When it is viewed in general, the architectural structure of the airport and the climatic conditions of the region are the most important factors that affect the energy consumption. There are many factors affecting energy consumption at the airports. These are [23, 26, 27]:

- Size of airport (conditioned spaces)
- Architecture of airport (compact, pier finger terminals, satellite terminals, and remote satellite terminals)
- Location—climate (climate)
- Operational hours
- Insulation level of terminal building
- HVAC system
- Utilizer behavior
- Energy management
- Airport maintenance level
- Capacity of aircraft maintenance facilities
- Daylight utilization
- Solar heating
- Traffic density
- Number of passenger
- Smooth operations of electrical and mechanical systems

While it is seen in Fig. 2.4 77% of total energy consumption at a sample airport is used in the terminal building, it is seen in Fig. 2.5 that the highest energy consumption in the terminal building is used for heating, cooling, and ventilation by 68%. To reduce energy consumption used for heating and cooling purposes, the following should be considered [28]:

- Application of optimum insulation thickness
- Using low thermal transmittance and high-performance windows
- Walls and roofs with reflective surfaces
- Prevents infiltration with sealing
- Passive solar design

Fig. 2.6 Organizational
structure at the airports
[29]

2.3.2   Management Organization of the Airport

There are different management organizational charts depending on the size and
capacity at the airports. Another aspect affecting the management structure is the
different management styles. While formal organizational structures are preferred at
the airports that adopt a centralist management approach, a lean organizational
structure is preferred at the airports with a flexible management approach. It is seen
that management levels are less and simple at the airports with flexible management
approach. However, there are also airports with mixed organizational structures.
When today's management structures at the airports are examined, simple organizational structures stand out as in Fig. 2.6 [29].

The energy management team at the airports should be included in the organizational structure shown in Fig. 2.6. For an effective energy management, the energy
management team should be connected to the airport management and within the
same organizational chart. The energy management team and its number may vary
depending on the size and activities of the airport. However, in an efficient energy
management system, the energy management team should meet the minimum requirements, and also they should be established taking into account the requirements.

2.3.3   Energy Management Organizational Structure

As it is seen in Fig. 2.7, the energy management team should be included in the
management organization structure at the airports, and the management representative should be connected to the top manager. The structure shown in Fig. 2.7

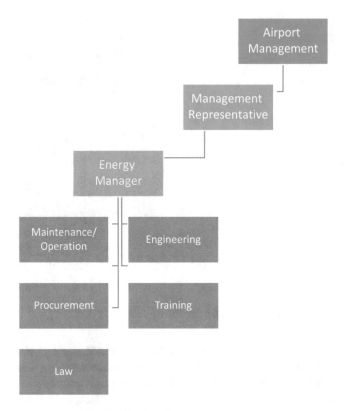

Fig. 2.7 Energy management unit at the airports

represents an exemplary energy management team for the airports. This structure may vary depending on the size and activities of the airports. When the relationship between energy and all aspects of sustainability (environmental, economic, and social) is taken into account, it makes it impossible to think of these two concepts as separate cases. For this reason, the energy efficiency activities and investments to be made energy efficiency, especially energy issue at the airports, should be environmentally, economically, and socially sustainable.

As it is seen in Fig. 2.7, the issue of energy management at the airports should be systematically addressed and represented at the top management level. An energy manager and an energy management team attached to the management representative should be appointed.

It has been demonstrated by many studies that energy consumption has direct environmental economic and social impacts. In this respect, while energy performance is continuously improved in energy management applications, it is recommended to create and implement optimum action plans economically and environmentally. It is an important issue for sustainable airport applications to take into account all effects (economic, social, and environmental) caused by energy consumption in the energy management logic to be created at the airports. The following options should be taken into consideration while creating action plans.

- Energy efficiency
- Greenhouse gas emission potential
- Human health
- Available resources
- Ecosystem damage
- Most environmentalist action plans
- Most economic action plans

2.3.4   Energy Audit

The purpose of energy audit is to determine the opportunities to decrease energy consumption and costs. After the energy manager is appointed, this person is provided with the necessary support to develop an adequate energy management program. The first step that the energy manager should take is to perform energy audits. The energy audit, which is also called energy analysis or energy review, is one of the most effective methods used in determining the opportunities to decrease energy consumption and costs. The objectives of performing energy studies can be listed as follows [30]:

- Determining the energy use and costs.
- Determining energy-inefficient equipment and processes.
- Identify new equipment and alternatives to reduce energy consumption and costs.
- Performing the economic and engineering analysis of these alternatives and to determine the most effective solutions.

American Society of Heating, Refrigerating and Air-Conditioning Engineers (ASHRAE) has formed three energy audit levels for commercial buildings. These are [31]:

ASHRAE Level 1: The level 1 audit is a quick method used to define energy efficiency opportunities by analyzing the energy systems and the past energy consumption data. The level 1 audits are used to determine the problematic points related to energy, to identify the measures with low cost or without cost, and to take the measures related to them in a short time. The results are used to determine the priorities in level 2 and level 3 energy audits.

ASHRAE Level 2: The level 2 audits are more comprehensive, and these studies include obtaining more detailed data from the past, collection of data with extensive field visits, and the determination and implementation of the opportunities through economic analyses. This stage includes the determination of low- and medium-cost action plans.

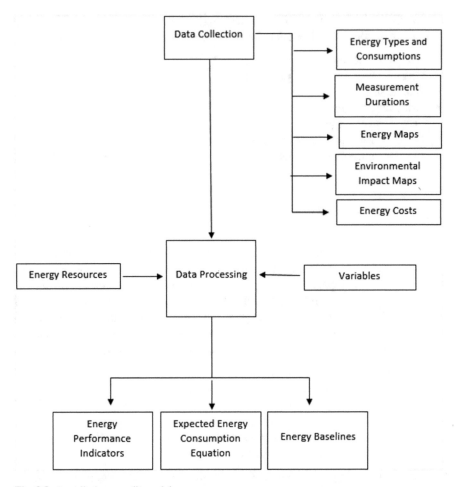

Fig. 2.8 A preliminary audit model

ASHRAE Level 3: The level 3 audits include more detailed examination and analysis of energy systems. More detailed data collection and the analysis of energy-consuming processes and equipment are performed in this stage. The economic and engineering analyses of the investments requiring high costs for the improvement of energy performance are performed in this stage.

2.3.4.1 Preliminary Audits

Preliminary audits are necessary processes to obtain energy costs, EnPIs, energy baselines, and expected energy consumption equations by using energy consumption data and variables that are thought to affect energy consumption. A preliminary audit model to be used in energy management system applications is presented in Fig. 2.8.

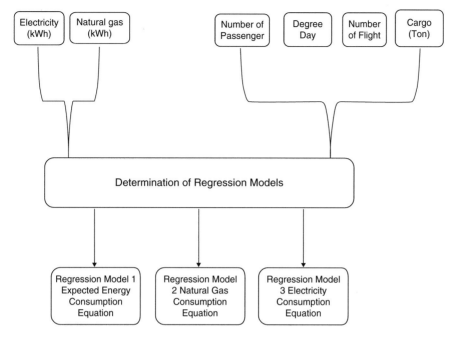

Fig. 2.9 Energy sources and variables used in the regression model

Energy costs and energy maps are generated by collecting the past data on energy consumption and types at the airports. The purpose of determining energy maps and their costs is to determine where and for what purpose energy is consumed and their costs. The variables affecting energy consumption at the airports can be determined by regression analysis. The variables that can be used at the airports are presented in Fig. 2.9. With the help of regression analysis, with which variables and to what extent energy consumption at the airport changes can be mathematically modeled. All variables that may affect energy consumption can be analyzed, and mathematical models related to them can also be determined.

2.3.4.2   Detailed Energy Audits

Detailed audits include the determination of energy inefficient points with the help of energy efficiency measuring devices, the establishment of action plans related to them, and the economic, engineering, and environmental analyses of these action plans. Furthermore, the determination of the legal and other requirements that should be complied with respect to structure and energy at the airports, the comparison of current situation with legal requirements, and taking necessary measures are also performed in this stage. The detailed audit model proposed for sustainable airports is presented in Fig. 2.10. The purpose of the detailed audits is to determine

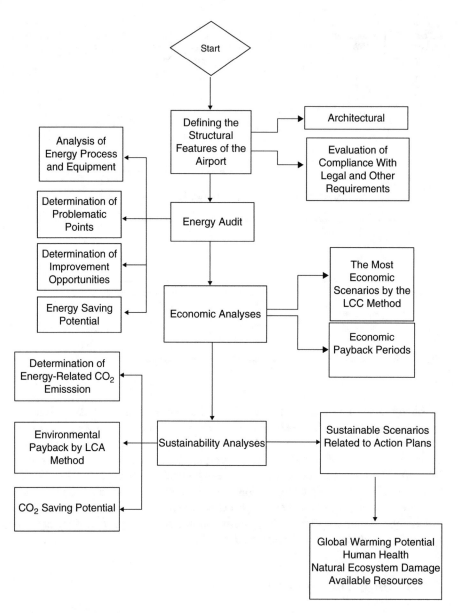

Fig. 2.10 Detailed study model

the energy inefficient points and to develop action plans related to them. The proposed method for developing action plans at sustainable airports is to determine the most economic and environmentalist scenarios using the LCC and LCA methods. In sustainable energy management practices at the airports, it is aimed to continuously improve both energy and environmental performance.

Detailed energy audits are performed using energy efficiency measuring devices. Action plans are developed for the specified problem points, and their economic, engineering, and environmental analyses can be performed.

In conclusion, energy costs, environmental impacts caused by energy consumption, and the legal requirements that need to be followed require the use of energy more efficiently at the airports. There are many factors affecting energy consumption at the airports. Excessive energy-consuming systems and equipment bring along effective energy management need. Energy issue at sustainable airports should be addressed systematically and holistically. Therefore, energy management at the airports should be represented at the top management level, and ISO 50001 standards, which are international standards, should be established and implemented at the airports. All stages of ISO 50001 should be considered as an opportunity to improve the energy and environmental performance of the airports. The fact that the employees working in the EnMS are experienced and well informed and work voluntarily is very important for an improved energy performance. The improvement opportunities can be determined by performing preliminary and detailed energy audits which are the most important stages of sustainable energy management practices at the airports. Nowadays, it has become a necessity to develop the action plans to be performed after the opportunities are determined by taking into account the principles of sustainable development (economic, social, and environmental).

References

1. Kılkış B, Kılkış Ş (2017) New exergy metrics for energy, environment, and economy nexus and optimum design model for nearly-zero exergy airport (nZEXAP) systems. Energy 140:1329–1349
2. Yüksel I (2010) Energy production and sustainable energy policies in Turkey. Renew Energy 35(7):1469–1476
3. Ozturk M, Yuksel YE (2016) Energy structure of Turkey for sustainable development. Renew Sust Energ Rev 53:1259–1272
4. Morvay Z, Gvozdenac D (2008) Applied industrial energy and environmental management, John Wiley & Sons, Chichester, United Kingdom
5. Bernstein L, Bosch P, Canziani O, Chen Z, Christ R, Davidson O, Kundzewicz Z (2007) Climate change 2007: Synthesis report. Contribution of Working Groups I, II and III to the fourth assessment report of the Intergovernmental Panel on Climate Change. IPCC, Geneva
6. ICAO, Aircraft engine emissions. https://www.icao.int/environmental-protection/Pages/aircraft-engine-emissions.aspx. Accessed 10 Sep 2018
7. ACI (2011) Airport carbon accreditation annual report. https://www.airportcarbonaccreditation.org/component/downloads/downloads/30.html. Accessed 10 Sep 2018
8. UNIDO (2015) Practical guide for implementing an energy management system. United Nations Industrial Development Organization, Vienna
9. Ortega Alba S, Manana M (2016) Energy research in airports: a review. Energies 9(5):349
10. Sogut MZ, Uysal MP, Gazibey Y, Hepbasli A (2016) Concept mapping of sustainable energy management for a holistic approach of energy strategies. Int J Global Warm 10(1/2/3)
11. McLaughlin L (2015) ISO 50001: energy management systems: a practical guide for SMEs. ISO, Geneva

12. Kanneganti H, Gopalakrishnan B, Crowe E, Al-Shebeeb O, Yelamanchi T, Nimbarte A, Abolhassani A (2017) Specification of energy assessment methodologies to satisfy ISO 50001 energy management standard. Sustain Energy Technol Assess 23:121–135

13. Ramamoorthy K (2012) A structured approach for facilitating the implementation of ISO 50001 standard in manufacturing industry. West Virginia University, Morgantown

14. ISO (International Organization for Standardization) (2011) ISO 50001 energy management systems—requirements with guidance for use. ISO Central Secretariat, Geneva

15. ISO 500001 – win the energy challenge with ISO 50001 http://www.segingenieria.com/admin/uploaded/archivos/iso_50001.pdf. Accessed 10 Sept 2018

16. TSE (2013) TS EN ISO 50001 Enerji Yönetim Sistemleri- Şartlar ve Kullanım İçin Kılavuz. Ankara (*in Turkish*)

17. Howell MT (2014) Effective implementation of an ISO 50001 energy management system (EnMS). ASQ Quality Press, Milwaukee

18. T.C. Resmi Gazete 27.10.2011. http://www.resmigazete.gov.tr/eskiler/2011/10/20111027-5.htm. Accessed 10 Sept 2018 (*in Turkish*)

19. Rietbergen MG, Blok K (2010) Setting SMART targets for industrial energy use and industrial energy efficiency. Energy Policy 38(8):4339–4354

20. Eccleston CH, March F, Cohen T (2011) Inside energy: Developing and managing an ISO 50001 energy management system. CRC Press

21. Thumann A, Mehta DP (2013) Handbook of energy engineering, 7th edn. The Fairmont Press, Inc., Lilburn

22. Gopalakrishnan B, Ramamoorthy K, Crowe E, Chaudhari S, Latif H (2014) A structured approach for facilitating the implementation of ISO 50001 standard in the manufacturing sector. Sustain Energy Technol Assess 7:154–165

23. Costa A, Blanes LM, Donnelly C, Keane MM (2012) Review of EU airport energy interests and priorities with respect to ICT, energy efficiency and enhanced building operation

24. Uysal MP, Sogut MZ (2017) An integrated research for architecture-based energy management in sustainable airports. Energy 140:1387–1397

25. ACI (2014) Airport energy efficiency and management. http://www.aci-asiapac.aero/services/main/17/upload/service/17/self/55cc67d1e0443.pdf. Accessed 10 Sept 2018

26. Akyuz MK, Altuntas O, Sogut MZ (2017) Economic and environmental optimization of an airport terminal building's wall and roof insulation. Sustainability 9(10):1849

27. Balaras CA, Dascalaki E, Gaglia A, Droutsa K (2003) Energy conservation potential, HVAC installations and operational issues in Hellenic airports. Energ Buildings 35(11):1105–1120

28. Dulac J, LaFrance M, Trudeau N, Yamada H (2013) Transition to sustainable buildings: strategies and opportunities to 2050. International Energy Agency, Paris

29. Şengür Y Yönetsel Sistem Olarak Havaalanı. F. Şengür (Editör), Havaalanı Yönetimi içinde (s. 30–59). Eskişehir: Anadolu Üniversitesi Açıköğretim Yayınları (2016.) (*in Turkish*)

30. Capehart BL, Turner WC, Kennedy WJ (2012) Guide to energy management, 7th edn. The Fairmont Press, Inc., Lilburn

31. ASHRAE (2015) ASHRAE handbook. Heating, ventilating, and air-conditioning applications. American Society of Heating Refrigerating and Air-Conditioning Engineers, Atlanta

Chapter 3
Risk Methodology to Assess and Control Aircraft Noise Impact in Vicinity of the Airports

Oleksandr Zaporozhets and Boris Blyukher

Analytical modelling of system safety and risk methodology is used to describe the relationships between the causes, dangers and effects of aircraft noise impact in various scenarios of aviation system development. Risk is assessed according to the hazard identification, associated with the probability of adverse events and their consequences, but the main emphasize is done on noise annoyance. The approach uses two types of risk in principle: individual and societal. A number of events must occur if a main stressor should take place with conditional probability of their realization. Dose-response function is applied to estimate the damage in an exposed receptor, and mathematically it gives a relationship between the intensity of the stressor and the effects in the exposed receptor. Framework for risk assessment and reduction concerning the control of a number of annoyed people by noise is proposed also. Hazard, vulnerability and coping capacity are interpreted in accordance with new requirements of the standard ISO 31000:2018.

3.1 Introduction

Environmental health concerns all the factors, circumstances and conditions in environment (in particular – in surroundings of humans) that can affect on human health and well-being [1]. Environmental noise is an obvious example of unwanted technological and social outcomes in continuous human development, with health and behavioural aspects for population under its impact. Throughout the world people

O. Zaporozhets (✉)
Nationasl Aviation University, Institute of Environmental Safety, Kyiv, Ukraine
e-mail: zap@nau.edu.ua

B. Blyukher
Indiana State University, Terre Haute, IN, USA
e-mail: Boris.Blyukher@indstate.edu.us

© Springer Nature Switzerland AG 2019
T. H. Karakoc et al. (eds.), *Sustainable Aviation*,
https://doi.org/10.1007/978-3-030-14195-0_3

perceive noise as an important issue that affects human health and well-being. Particularly in Europe one-third reported problems with noise (ranging between 14 and 51% in particular states) are observed in urban conglomerates mainly and their large proportions connected with noise annoyance [2]. Among other sources an aircraft noise can also be a substantial source of population annoyance – this is the most objective measure of this health outcome.

The aircraft noise is one of the most important local impact factors on environment (somewhere the single dominant factor) arising from airport operations, which, unless managed effectively, has the potential to constrain the ability of airports to grow in response to demand and hence to limit the social and economic benefits that future growth could bring. Together with these and various other social, safety and economic problems, including a number of specific environmental issues, aircraft noise has the potential to multiplying constrain the operation and growth of the airports and air traffic in general [3], Fig. 3.1.

In particular, the optimisation can be used to search for cost-minimal balances of controls of all the factors under consideration over the various scenarios in aviation sector development (which considered as a system [4, 5]) that simultaneously try to achieve user-specified targets for human health impacts (e.g. expressed in terms of reduced life expectancy in accordance with WHO guidelines for these factors – long-term exposure to noise pollution from any type of traffic may reduce life expectancy, a new study contends [6]), ecosystems protection, maximum allowed violations of WHO guideline values [7], etc. (Fig. 3.1).

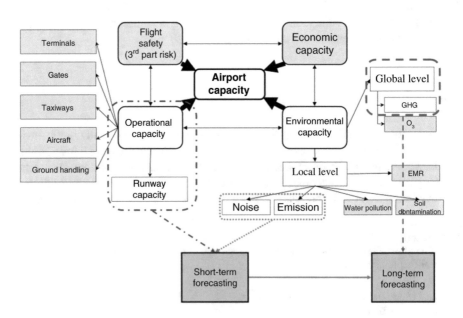

Fig. 3.1 Airport interdependency and capacity analysis scheme [3]

The International Civil Aviation Organization (ICAO), retaining its own key function, is conscious of and will continue to address the adverse environmental impacts and will strive to [8]:

(a) Limit or reduce the number of people affected by significant aircraft noise.
(b) Limit or reduce the impact of aviation emissions on local air quality.
(c) Limit or reduce the impact of aviation greenhouse gas emissions on the global climate.

These goals are quite unbalanced between themselves, because for noise the main goal is defined directly as a "number of people affected by significant aircraft noise" – real value of evident consequences of noise impact on environment. While the ICAO emission goals "impact of aviation emissions on local air quality" and/or "impact of aviation greenhouse gas emissions on the global climate" are possible to be considered much wider, including the number of people affected by aviation emissions, but usually an inventory analysis is still used to describe both of them in any kind of environmental assessment, not exposure/impact analysis as for noise. Such misbalance may influence dramatically on any kind of environment impact assessments for any decision preparations made for environment protection management in aviation sector, especially if an optimization task should be formulated and solved, similar to the approach, realized in [4, 5].

To find an approach for more balanced definition of the goals for environment protection from civil aviation impact, there is a risk methodology proposed. Implementing an al-hazards assessment approach (considering a number of factors of aviation impact on environment or even wider – e.g. a number of factors of hazard, including safety and security issues) that combines the both – as natural so as the man-made hazards (incorporating technological and biological hazards, not only usually used chemical and physical, that can have multiplying effects that exceed the boundaries of a system under consideration) and including all elements of possible risk – will require an integrated/collaborative approach across disciplines and sectors as well as inter-science and technology cooperation [9]. In this paper the risk methodology is considered in connection with aircraft noise impact mostly, with emphasize on noise annoyance.

3.2 Risk Methodology in General

In accordance with the UN International Strategy for Disaster Reduction (UNISDR) terminology [10], the risk is defined "as the probability of harmful consequences, or expected losses, resulting from interactions between natural or human-induced hazards and vulnerable conditions". The losses are considered directly for humans as their deaths and injuries and the damages for their property and livelihoods and for economic activity disrupted or damaged environment.

Under the current standard [11], the definition of "risk" is no longer a "chance or probability of loss" but an "effect of uncertainty on objectives". The purpose of risk assessment is to provide evidence-based information and analysis to make informed decisions on how to deal with any of risks and how to choose the best between alternative options. Principal benefits of a performing risk assessment include a wide set of positive outcomes [11]:

– Providing objective information (calculated or measured, or even qualitative) for decision-makers.
– An understanding of the risk and its potential impact upon objectives.
– Identifying, analysing and evaluating risks and determining the need for their treatment.
– The quantification or ranking of risks, especially for different types of the stressors like noise, emission/pollution in air, water, soil, etc.
– Contributing to the understanding of risks (correct discription of the stressor exposure and vulnerability of the receptor to this stressor) in order to assist in the selection of treatment/prevention options.
– Identification of the important contributors – direct and indirect – to risks and weak links in systems and organizations.
– Comparison of risks in alternative systems, technologies or approaches for the choice of the preferable between them - with smalest stressor and vulnerability of the receptor to the stressor.
– Identification and communication of risks and uncertainties: uncertainty may influence dramatically the decision making - the choice between the alternatives.
– Assisting with establishing priorities for health and safety.
– Providing information that will help evaluate the tolerability of the risk when compared with predefined criteria and others.

In general the risk (R) and hazard (H) ratio can formally be expressed in simple form as:

$$R = f\left(H \times E\right),\tag{3.1}$$

where H, hazard, is in our case noise generated by flight event (aircraft in flight) of specific value, which may lead to a number of effects (for humans, ecosystems, etc.); E, type and value of exposure of the hazard (aircraft noise in particular) on subject of impact (e.g. on population); and f, function of their interdependence.

Exposure is a phenomenon of the *stressor* (synonymous with the terms *agent*, *factor* – any physical, chemical or biological entity (a phenomenon, object, substance, etc.,) which may cause unacceptable response, usually called *damage*) and has contact with the *receptor* and may be defined as "the people, property, systems or other elements (elements-at-risk) present in hazard zones (areas, where the hazard may spread with valuable level) that are thereby subject to potential losses" [10]. *Receptor* is an entity that is exposed by the stressor. The effect for receptor is determined by the potential consequence to it, most often in the form of the damage caused by the hazard that exists at particular state of the system under consideration. Usually an exposure E of the physical, chemical or biological stressor is a function of its dose D and time t of interconnection between stressor and receptor.

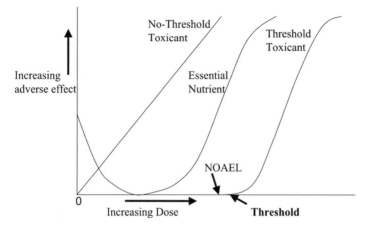

Fig. 3.2 Dose-effect curves for various types of response to hazard: NOAEL, no-observed-adverse-effect level

A dose-response curve (Fig. 3.2) records the percentage of a population showing a given quantal (all or nothing) response such as death or illness when each individual member of the population is subjected to the same dose of toxicant (reflecting a given exposure of chemical or biological stressor). Human response to noise, depending on type of possible effect in this case, is usually described by the curve, which is quite similar with essential nutrient one.

This simple conceptual dependence between risk and hazard in Eq. (3.1) does not consider the contribution of vulnerability of the elements-at-risk to the hazard under consideration – "the conditions determined by physical, social, economic and environmental factors or processes, which increase the *susceptibility* of a community to the impact of hazards" [10]. Elements-at-risk has a certain level of vulnerability usually. In general case the vulnerability "describes such *characteristics* and *circumstances* of a community, system or asset under consideration that make them susceptible to the damaging effects of a hazard" [10]. Relating to a number of inter-related conditions (they can be generally classified as shown in Table 3.1), vulnerability may increase the susceptibility of a community to the impact of any hazards under consideration [12].

Risk assessment is concerned with determining those factors which are especially determining the likelihood of unacceptable harmful exposure. Among vulnerability properties of the population, under the risk of noise impact is a number of acoustic (aircraft type, especially type of the engine in aircraft power plant, fleet composition in specific airport under consideration, flight route distribution around the airport, their respective distribution of the flight traffic over given time period of observation) and non-acoustic drivers (personal noise sensitivity, attitude towards the noise source, performed activities at the moment, etc.). For example, fleet composition in contrary to number of flights, which directly defines number of noise events and a value of noise exposure on elements-at-risk, may be considered as a vulnerability effect, for example, in case with a number of turboprops in a fleet

Table 3.1 General classification of vulnerability [13]

	Human–Social	Physical	Economic	Cultural Environmental
Direct losses	Fatalities Injuries Loss of income or employment Homelessness	Structural damage or collapse to buildings Non-structural damage and damage to contents Structural damage infrastructure	Interruption of business due to damage to buildings and infrastructure Loss of productive workforce through fatalities, injuries and relief efforts Capital costs of response and relief	Sedimentation Pollution Endangered species Destruction of ecological zones Destruction of cultural heritage
Indirect losses	Diseases Permanent disability Psychological impact Loss of social cohesion due to disruption of community Political unrest	Progressive deterioration of damaged buildings and infrastructure which are not repaired	Economic losses due to short term Long-term economic losses Insurance losses weakening the insurance market Less investments Capital costs of repair Reduction in tourism	Loss of biodiversity Loss of cultural diversity

(low-frequency noise is dominant in their individual exposure levels) instead of more used turbofans and turbojets in current fleets everywhere.

In relation to hazard (H), vulnerability (V) and amount of elements-at-risk ($A_{elements\text{-}at\text{-}risk}$) (or consequences to them from hazard impact), the risk (R) can be presented conceptually with the following basic equation:

$$R = H * V * A_{elements-at-risk} \qquad (3.2)$$

or taking into account the capacity (C_c) (opposite term to vulnerability) to cope the hazard consequences [14]:

$$R = H * V * A_{elements-at-risk} / C_c. \qquad (3.3)$$

All these formulas (3.1)–(3.3) would not be taken literally as a mathematical formula in most of the cases but rather a model to demonstrate a concept. The first part of the formula for risk, hazard (or somewhere threat) × vulnerability, can also be looked at as a probability [14]. This likelihood is a rough measure that describes the chances a given vulnerability will be discovered and used by a hazard actor. The last part of the formula ($A_{elements-at-risk}$) describes the consequences, or impact itself, of an impacting attack by hazard actor. The combination of the likelihood and the impact describes the severity of the risk, Fig. 3.3.

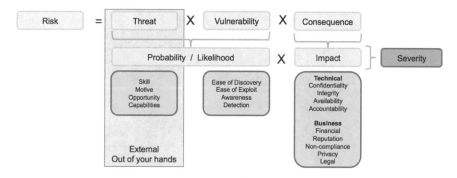

Fig. 3.3 Scheme for risk calculation

Recently, various disciplines investigated the concept of interaction of danger, vulnerability and coping capacity for risk assessment and control. Its conceptual risk assessment formula has changed over time as follows [14]:

$$R = f\left[H\left(V,C_c\right) \cdot V\left(H,C_c\right) / C_c\left(H,V\right)\right], \tag{3.4}$$

where a number of complex interactions between attributes hazard (H), vulnerability (V) and capacity (C_c) are considered for any possible kind of elements-at-risk. For example, the response of the human ear to acoustic spectral frequency of noise event may be considered as vulnerability property of the humans under the risk of noise impact or as it is used currently – to correct the noise exposure level on a value of human ear response, in such way to change the impacting exposure (or hazard) of this disturbing noise including human susceptibility to sound frequency. If the humans will be provided with ear plugs to prevent them from disturbing noise (e.g. during their sleep) or if the disturbing noise source will be outside of the sleeping room and a simply closed windows may provide less vulnerable conditions – once again a complex interdependency between H, V and C_c will take place (e.g. closed windows in a room may be considered as coping capacity of the location for noise protection purposes or as vulnerability conditions of this location) – their account on noise impact will be quite complicated. Any kind of community engagement may influence a problem of aircraft noise impact assessment dramatically due to such complex interdependency between H, V and C_c (Fig. 3.4).

In general case the severity of the hazard impacts depends on the level of exposure inside the affected area; if it is absent, an impact is absent also. But evidence exists that a number of risk have increased worldwide not only due to increases in hazard exposure of population and/or its assets but the vulnerability is also fundamental to our understanding of risk [9]. For example, increasing aircraft noise exposure has been the major cause of long-term increases in economic losses in aviation sector. There have been localized reductions in vulnerability as a result of, for instance, better building standards and compliance (sound insulation measures in buildings, noise screens outside, etc.), but on the other side, there are many instances

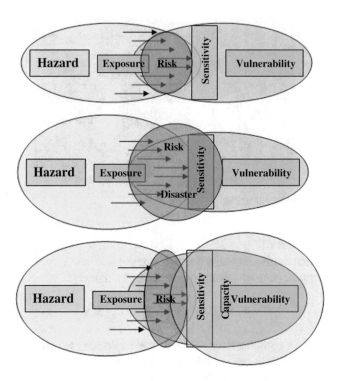

Fig. 3.4 Scheme for hazard, vulnerability and coping capacity interinfluence in risk assessment and control [14]

of increased vulnerability, particularly due to huge population densities in large urban centres and in number of developed and developing countries. Only due to increased vulnerability a significant increase was observed in human annoyance at a given level of aircraft noise exposure over the last years – decade or two. Crucial evidence that annoyance measured in European airports (the more recent studies in airports of Manchester, Paris, Amsterdam, Frankfurt, all of them during the last decade) is much higher dependent from the noise indices [15], the explicit disparateness in value of around 5–6 dBA between the average trend of all of these more recent studies and the older data executed in standards; it means that currently a more high number of annoyed people may be observed in acoustic conditions which were considered not so serious a decade or two before (Fig. 3.5). These results are of highest importance to the applicability of executed standards of exposure-annoyance relationships for aircraft noise and provide a basis for decisions on whether these need to be updated or a new approach for their assessment.

But the given Eqs. (3.1)–(3.4) are not representing a concept only; sometimes they can also be used for pure calculation (e.g. to quantify risk from geo-distributed hazards using the spatial data in a GIS [13]).

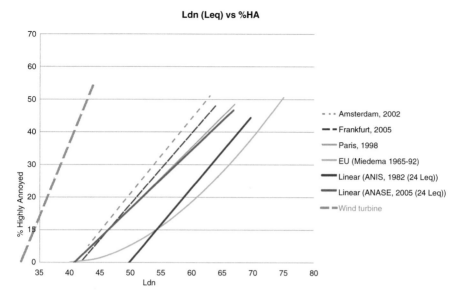

Fig. 3.5 Aircraft noise annoyance curves for a number of studies in EU [15], combined with known wind turbine noise annoyance curve

Mathematically the risk is proportional to a measure for the probability (P) of an event (frequency, likelihood), creating a hazard, and the consequences (C) of an event (impact, effect on objectives) or hazard consequences (Fig. 3.2):

$$R = P * C. \tag{3.5}$$

For *individual* risk this basic condition may be expressed by the formula [16]:

$$R = P_f * P_{d/f}, \tag{3.6}$$

where P_f is the probability of harmful event (e.g. aircraft accident) and $P_{d/f}$, the likelihood of the consequences (effect or damage), particularly the fatal consequences caused to individuals in the absence of protection from (or resistance to) a danger (Fig. 3.5).

Individual risk (R_i) is an averaged over the year probability of death, injury and illness for an individual who may be located (lives or performs any kind of activities) near the source of hazard (e.g. airport or power plant or any other critical object), as a result of hazard existence/occurrence/exposure (e.g. airplane crash into power plant) and regardless of the presence of the people at all. The purpose of R_i estimating is to ensure that individuals, who may be affected by a threat from critical object (source of hazard), are not exposed to excessive risks. It is a characteristic of the source of hazard or property of lands around this object (which is location specific); thus R_i may be shown by contours around the object on a map and to be used further for land use planning and corresponding zoning purposes to control the impact of this hazard on population.

Also a *societal risk* (R_S) may be used for assessment, which represents the risk to a society, or community or simply group of people. It is an annual probability that N or more people may die, being injured and/or ill due to the danger occurrence/ exposure. R_S is not person and location specific. Because societal risk is multidimensional usually, it is difficult to apply to the task of risk reduction. It is therefore reasonable to examine both R_S and R_i to achieve a full-risk picture to be effective with risk management in the following steps.

Severity of a hazard (risk of consequences of a danger, Fig. 3.2) is combined with an estimate of its probability (or consequence). First needs to be determined, how often there may be a danger. Usually a function of probability combinations of causes (factors) should be considered. Then a likelihood of the worst state of the system must be assessed. This evaluation can also be quantitative or qualitative.

Calculation of individual risk (R_i) with formula (3.6) – the multiplication of the probability of hazard event and the damage given by this event – provides logic that R_i can never become larger than the probability of hazard event inside a system, because the damage fraction is never larger than 1. By integrating the individual risk (R_i) and the population density m, the expected value of the number of people with damages for their health $E(N)$ inside population N can be determined:

$$E(N) = \iint_A R_i(x,y) m(x,y) dxdy, \tag{3.7}$$

where all the contributing values are defined at location (x,y) and number of damaged people inside area A per year. The total number of people exposed (N_{EXP}) to a certain hazard event is usually higher and can be found by integrating the population density $m(x,y)$ over the exposed by hazard area A:

$$N_{EXP} = \iint_A m(x,y) dxdy. \tag{3.8}$$

In more general form, probability of harmful event P_f may be divided to the probability of scenario p_{Sc}, leading to such event, and the probability of hazard exposure p_{Ex} due to this scenario:

$$P_f = p_{Sc} p_{Ex}. \tag{3.9}$$

The effects are usually described in terms of various types of damage k (e.g. fatality, injury, physical damage, environmental losses, loss of income, etc. depending what are the elements-at-risk) and their vulnerability v_k (e.g. describing a third-party risk around the airport in case of aircraft accident, a person's vulnerability can be defined as mortality):

$$P_{d/f} = k * v_k. \tag{3.10}$$

Simple classification of different types of consequences due to technological event (or accident) is given in Table 3.2. This classification of damages covers tangible and intangible types, depending on whether or not these losses can be estimated in

Table 3.2 General classification of damage, based on [17]

Damage	Tangible	Intangible
Direct	Residences	Fatalities
	Airport facilities and inventory	Injuries and illnesses
	Vehicles	Animals
	Agriculture	Utilities and communication
	Infrastructure and other public facilities	Historical and cultural
	Business interruption (inside affected area)	losses
	Evacuation and rescue operations	Environmental losses
	Cleanup costs	
Indirect	Damage for business outside affected area	Societal disruption
	Substitution of business/production outside affected area	Damage to government
	Temporary housing of evacuees	

monetary values. Another dissimilarity is found between the direct damage, for example, caused by physical contact with aircraft crash just on site of the accident, and damage indirectly following from the crash (fire, air pollution outside the accident site and so on). Indirect effect can be defined also if damage may occur outside the affected area or/and over the delayed period of time [17]. For example, any kind of business can lose supply and demand during the time from the affected area, for example, the demand to have a residence in noisy area.

For aircraft noise any flight event is leading to scenario of noise impact, $p_{Sc} = 1$; the same is valid for aircraft engine emission and air pollution, but the probability of hazard exposure p_{Ex} due to any scenario is dependent of specific location of point of control relatively the flight path – people are impelled to complain when some burden factor in the environment gives rise to any effect and when this stressor reaches a lower limit value (Table 3.3). Aircraft noise exposure can lead to more than one effect, and the community impacts (usually health effects, which can be chronic) depend on multiple effects (also shown in Table 3.3) [18]: the primary recognized health consequences of community noise exposure are the sleep disturbance during night-time and annoyance during composite daytime, and anywhere due to vulnerability aspects, the cardiovascular disease and cognitive impairment in children also contribute [7]. Efforts to reduce exposure should primarily reduce annoyance and sleep disturbance, improve learning conditions for children and lower the prevalence of cardiovascular risk factors and cardiovascular disease [18] – they usually have different coping capacities for all these types of health consequences. Evidence is increasing to support preventive measures separately to them, such as noise insulation, policy, guidelines and limit values.

If k is correspondent to noise annoyance effect, the likelihood of $P_{d/f}$ may be represented as a dependence of $HA\%$ from noise exposure E; currently L_{DN} (or its analogue L_{DEN}) is used as its metric because it is mostly correlated with noise annoyance of the population living under the noise impact around the airports (Fig. 3.6).

Risk assessment needs to be used in framework of its regulation (Fig. 3.7) [20]. To investigate the effects of hazards correctly, there are important factors of vulner-

Table 3.3 Critical limits, protection guides and threshold values for sleep disturbance and annoyance [19]

Effect	Evaluation criteria	Measure	Value	Indoor/outdoor
Sleep disturbance	Critical limit[a]	dB $L_{Amax\ 22\text{-}06\ hour}$	6 events at 60 dBA	Indoor
	Critical limit	$L_{Aeq\ 22\text{-}06\ hour}$	40	Indoor
	Protection guide[b]	dB $L_{Amax\ 22\text{-}06\ hour}$	13 events at 53 dBA	Indoor
	Protection guide	dB $L_{Amax\ 22\text{-}01\ hour}$	8 events at 56 dBA	Indoor
	Protection guide	dB $L_{Amax\ 01\text{-}06\ hour}$	5 events at 53 dBA	Indoor
	Protection guide	dB $L_{Aeq\ 22\text{-}06\ hour}$	35	Indoor
	Protection guide	dB $L_{Aeq\ 22\text{-}01\ hour}$	35	Indoor
	Protection guide	dB $L_{Aeq\ 01\text{-}06\ hour}$	32	Indoor
	Threshold value[c]	dB $L_{Amax\ 22\text{-}06\ hour}$	23 events at 40 dBA	Indoor
	Threshold value	dB $L_{Aeq\ 22\text{-}06\ hour}$	30	Indoor
High annoyance	Critical limit	dB $L_{Aeq\ 06\text{-}22\ hour}$	65	Outdoor
	Protection guide	dB $L_{Aeq\ 06\text{-}22\ hour}$	62	Outdoor
	Threshold value	dB $L_{Aeq\ 06\text{-}22\ hour}$	55	Outdoor
Chronic disease	Critical limit	dB $L_{Amax\ 06\text{-}22\ hour}$	19 events at 99 dBA	Outdoor
	Critical limit	dB $L_{Aeq\ 06\text{-}22\ hour}$	70	Outdoor
	Protection guide	dB $L_{Aeq\ 06\text{-}22\ hour}$	25 events at 90 dBA	Outdoor
	Protection guide	dB $L_{Aeq\ 06\text{-}22\ hour}$	65	Outdoor

[a]*Critical limits* – above these levels there is a risk of health effects, and such levels should only be tolerated as an exception for a limited time. Above these levels there is already a noise, so it is imperative that noise control measures should be introduced

[b]*Protection guides* – exposure below these levels should not induce adverse health effects in the average person, although sensitive groups may still be affected. These are the "central assessment values" above which action should be taken to reduce noise exposure

[c]*Threshold values* – inform about measurable physiological and psychological reactions to noise exposure where long-term adverse health effects are not expected. To increase quality of life, these values constitute a long-term goal

ability (mentioned before: physical, social, economic and environmental conditions and processes) that tend to increase the damage from the effects of the hazards impact on the person or society as a whole (Fig. 3.8). There is a necessary coping *capacity* – capabilities of a human, system, society and nature to confront the consequences of dangers and threats, i.e. resources are needed that may reduce the negative effects.

It was found that human response to noise is varying differently among environmental sound sources that are observed to have the same acoustic levels (Fig. 3.6b). Due to these and other differences, ISO 1996-1 [23] provides and describes a number of adjustments for sounds that have different characteristics. The term "rating level" is used to describe physical sound predictions to which these adjustments may have been added. On the basis of such rating levels, the long-term community response can be estimated, for example, an Eq. (D.1) in ISO 1996-1 [23] is the original Schultz interpolation curve (Fig. 3.6a), showing the portion of a community that may be assessed as highly annoyed by transportation noise sources in dependence to the long-term (up to the year averaging) day-night sound level. The equation is

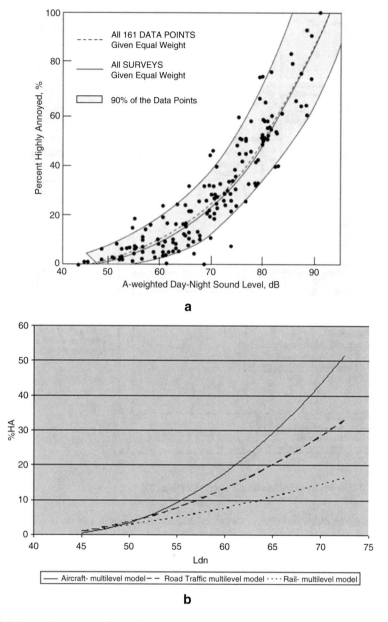

Fig. 3.6 Dose-effect curves for environmental noise – a portion of highly annoyed people in exposed by noise group correlated with day-night average sound level L_{DN} for (**a**) EPA dose-response relationship, developed by Schultz [21]; (**b**) three modes of transportation [22]

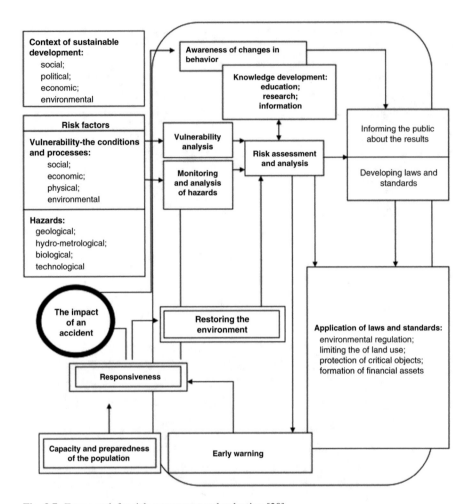

Fig. 3.7 Framework for risk assessment and reduction [20]

considered as not valid over short time periods and should not be used for assessments of community response during the "weekends, a single season, or "busy traffic days". Taking this in mind, it may be proposed to be considered as a normalized value of noise annoyance in population $L_{DN\ ISO}$, so any particular type of environmental noise (any transportation type like in Fig. 3.6b, or building construction, or wind turbine noise, etc.) $L_{DN\ s}$ should be assessed using specific noise source adjustment ΔL_s, which is character for environmental noise under consideration:

$$L_{DN\,s} = L_{DNISO} + \Delta L_s. \tag{3.11}$$

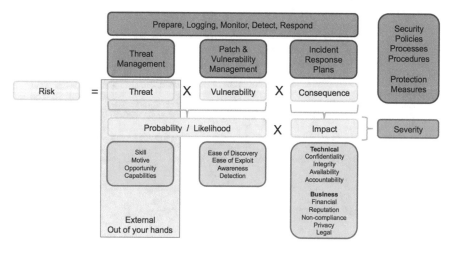

Fig. 3.8 Scheme for risk calculation and management

3.3 Aircraft Noise Annoyance

Noise pollution around airports is a major problem with regard to environment in modern societies, causing crucial negative effects on the communities in their vicinity (health and performance effects, health-care costs, properties depreciation, loss of operation by airports, etc.). Among other noise effects (Fig. 3.9), *noise annoyance* is the dominant subjective response to noise and the great reason of complaints in airports and, therefore, of concern for airport authorities. The combination of growing airport development around the world and emerging public concern about aircraft noise disturbance and annoyance continues to rise, so considerable mitigation efforts by governmental administrations and the civil aviation industry are expected to cope them.

There are also arguments that environmental noise can be considered, like in occupational noise case, as a risk factor for health, never mind that hearing impairment is not usual outcome of environmental noise impact, including aviation noise impact. A brief analysis of these arguments for assessing the risk of noise effects on health is provided in Table 3.4, including the following effects [24]: psychosocial such as annoyance and other subjective assessments of human well-being and quality of life, mental health, sleep which can be both psychological and physical, physical health such as hearing loss, and stress-related health which may include psychological, behavioural, somatic and physical symptoms.

Aircraft noise annoyance is still an increasing problem, especially in the densely populated areas like urban conglomerations and without procedures (usually systematic approach should be provided) reducing population's annoyance to environmental noise, so with time it becomes more difficult to increase the aircraft traffic or to build new runways or other airport infrastructure to grow the airport

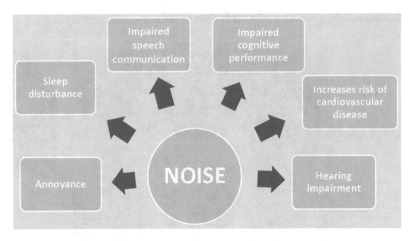

Fig. 3.9 Environmental noise effects

Table 3.4 Noise effects and conditions of noise exposure and receptor vulnerability (not full list of all, only the example of interconnections between the receptor, exposure and vulnerability conditions and type of effect)

Effect of noise impact	Receptor and their main effect symptoms	Exposure conditions	Vulnerability conditions
Aircraft noise annoyance	Behaviour responses are to be expected for children and adults	Annoyance responses in relation to noise exposure may be higher than predicted by the standard noise/annoyance curves, with effects observed for daytime (L_{Aeq16h}) and night-time ($L_{Aeq\,8h}$) exposure	Annoyance responses may increase in relation to operational changes; where populations become newly exposed to noise; where populations experience a step change in exposure; in response to early morning and evening flights
Risk for poor cardiovascular health outcomes	High blood pressure, heart attacks, stroke as well as with cardiovascular hospital admission and mortality	Increases in risk become important if a large population is exposed to aircraft noise	High blood pressure and cholesterol can be treated with medication to avoid more serious cardiovascular disease progression
Self-reported sleep disturbance	Sleep interruption and changes in sleep structure	Populations exposed to night-time noise could benefit from insulation of their home. It may also be beneficial to consider the use of curfews for night-noise flights	Not be restricted only to the night-time period, possible to be observed during the evening and early morning
Psychological ill-health associated with decreased quality of life	Hyperactivity symptoms observed for children are small and unlikely to be of clinical significance in the population exposed	Populations exposed to daytime noise during human activities and night-time noise during sleep	Populations exposed to noise could benefit from insulation of their activity and residential buildings

capacity. Surveys assessing baseline annoyance or/and monitoring of human annoyance responses at different times of the day prior to the development of the new runway would be useful for comparative purposes. Such monitoring would be helpful to identify any increases in annoyance related to operational decisions in and around the airport. According to the WHO, an outdoor noise level exceeding 55 dB L_{DN} is considered to be "seriously annoying" [7] (as it is shown in Table 3.3). From this point airport capacity will be limited huge from this noise annoyance acceptability level.

In order to represent the greater annoyance caused by a sound intrusion at night, the day-night sound level has been derived. It supposes that the equivalent sound level (L_N) occurring between 22.00 and 7.00 [23] should be augmented by 10 dBA before being combined with the equivalent sound level (L_D) for the period 7.00 to 22.00 to give the day-night level:

$$L_{dn} = 10 \log\left\{\frac{1}{24}\left[15\cdot10^{0.1L_D} + 9\cdot10^{0.1(L_N+10)}\right]\right\} \qquad (3.12)$$

The term annoyance is used to describe negative reactions to noise such as disturbance, irritation, dissatisfaction and nuisance [25, 26]. Annoyance from any source represents a diminished state of well-being, and noise is often referred to as the stressor that is implicated in a variety of responses (Fig. 3.10). In their "evidence

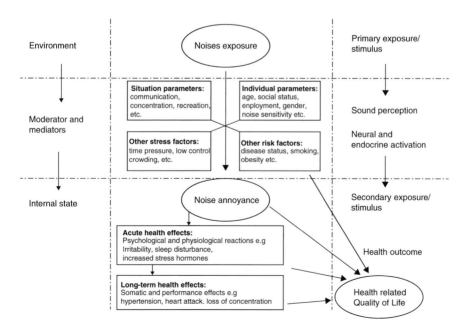

Fig. 3.10 Conceptual model of nonauditory effects of environmental noise and noise annoyance. (Reproduced from Ref. [27])

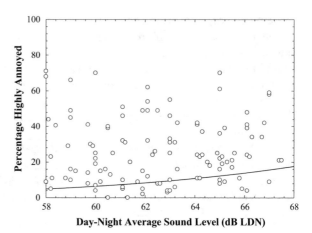

Fig. 3.11 Percentage of highly annoyed people at aircraft noise plotted as a function of noise exposure – day-night noise index L_{DN}

review of annoyance" paper, the WHO described the complex annoyance response to noise as comprising three main elements.

Scrutiny of Fig. 3.11 discloses that annoyance reactions to noise vary substantially in variety of possible cases of exposure and do not appear to be correlated with sound exposure level of noise only. What is more difficult to describe and measure is that there are many factors that give rise to annoyance and are the nonauditory effects of noise dealing with nuisance. There is no agreed approach to combine everything into an overall annoyance response currently, even if this were meaningful when taken out of the context of the many and varied social and economic factors that often have much greater health impacts. The solid curve in Fig. 3.11 is a portion of that presented in Fig. 3.3, while the scattered points represent results of real studies [28], any of these studies' results may vary substantially in noise (stressor) exposure and vulnerability conditions of the receptor. The shown curve is normalized, so a number of vulnerability adjustments, described further, should be used for correct evaluation of the percentage of highly annoyed people exposed to noise with level (index) L_{DN}. The results of the different studies – points around the normalized curve – are spread inside vulnerability corridor, which boundaries are quite close to the boundaries of the 90% of the data points in Fig. 3.6a and where the human annoyance to noise may be expected.

The equivalent sound level for the day and night in Eq. (3.11) in accordance with standard ISO 1996 [23] is the level of a steady sound, which, in a given period of time, would contain the same noise energy as the time-varying sound level one is describing (since the level at any location will generally vary with time). The concept of *equivalent sound level* is grounded on approach that the same amount of *noisiness* occurs if the same amount of acoustic energy is involved, so it is equal from a sound having a high level for a short period, as from a sound having a lower level but occurring for a long enough period. Most sounds vary irregularly in level, the aircraft noise flight event especially, and the derivation of L_{eq} demands an integration of the sound intensity on a continuous basis.

In mathematical terms, the A-weighted equivalent continuous sound pressure level is defined as:

$$L_{Aeq,T} = 10 \ \log \frac{\frac{1}{T} \int_{t_1}^{t_2} p_A^2(t) dt}{p_0^2} \tag{3.13}$$

where $p_A(t)$ is the A-weighted instantaneous sound pressure at running time t and p_0 is a reference value equal to 20 µPa or in more simple form:

$$L_{eq} = \frac{1}{T} \int_T^0 10^{0.1 L_A} \, dt \tag{3.14}$$

where L_A is the sound level, generally expressed in dBA, and T is the time period for which we are describing the sound. This definition in Eqs. (3.13) and (3.14) however takes into account the response of the human ear to noise of various frequencies and intensities and compensates these effects by electrical weighting network (filter of type "A", which describes a human ear response) incorporated in the measuring system.

The WHO Community Guidelines [7] formulate a *precautionary principle* to environmental noise effects on human health, and these guidelines are often considered by acousticians and policy makers to be very difficult to achieve in real practice. The WHO Europe Night Noise Guidelines [29] state that the target for nocturnal noise exposure should be 40 dB L_{night} outside the building, which should protect the public in general as well as vulnerable groups such as the elderly, children and chronically ill from the effects of nocturnal noise exposure on their health. The Night Noise Guidelines [29] also recommend to maintain 55 dB L_{night} outside the building as an interim target, in such a way to realize a stepwise approach to the programme on noise protection.

Evidently the time of day plays a dominant role when humans indicate their noise annoyance [30]. At same noise levels (sound equivalent or exposure levels or noise indices), night-time annoyance is above daytime annoyance, the effect is observed for aircraft noise in any case, while in the case of road traffic noise, it is absent, no difference is observed in annoyance between day and night response of the people. Vice versa for railway traffic, people are less annoyed by its noise during the night-time than during the day. The Swiss Noise Study 2000 demonstrated [30] that people feel highly annoyed especially in the morning, around noon and in the evening. It has sometimes been suggested that the 24 hours should be divided into three segments (day, evening and night) rather than two, with an evening (4 hours between 19:00 and 23:00 usually) weighting of perhaps 5 dBA. This has not been shown to be necessarily advantageous. It has also been suggested that the night-time weighting is not worth incorporating, in which case we are left with just L_{Aeq}.

Exposure covers a number of acoustic factors, which are first of all the maximum sound exposure levels, number of flights during the period of observation, usually during the day. Over the years, many attempts have been made to relate the percentage of respondents highly annoyed by a specific noise source to the day-night average noise exposure level, L_{DN}, or a similar indicator, e.g. L_{DEN} [31, 32]. One may assume that risk for population living around the airport to be highly annoyed by aircraft noise is defined by the day-night average noise exposure level. The standard ISO 1996: 2016 [23] has tables with % HA as a function of *DNL* and *DENL* for various transportation noise sources: they are quite different for aircraft (mostly impacting the people), railway and road traffic noise sources (Fig. 3.6b). Latest review prepared for WHO [26] still suggests the different relationships, particularly/especially for aircraft noise annoyance. A review by [33] confirms these data for aircraft noise also [34]. Results show also that if road and aviation noise are impacting jointly (the combined effects of transportation noise), the perception of the total noise annoyance was strongly determined by the sound source which was examined as more annoying (in this case aircraft noise).

But among the aircraft noise sources, anybody should recognize the following: aeroplane noise, helicopter noise, UAM/UAS noise, supersonic aircraft noise, etc. Never mind all of them are characterized by the same acoustic performances, the attitude towards the noise source and the correspondent impact the humans realized in a different way, so the annoyance outcomes are different for them also, like the difference exists for various transportation noise sources (Fig. 3.6b).

In any case, if doing nothing with annoyance management, the future scenarios for air transportation are looking unsatisfactorily, because existing results of the studies of noise annoyance in other sectors, different from aviation and having the same (e.g. with turboprops) or quite similar acoustical properties, for example, wind turbine noise (Fig. 3.12), show the much higher annoyance than existing aircraft noise studies (Figs. 3.5 and 3.6).

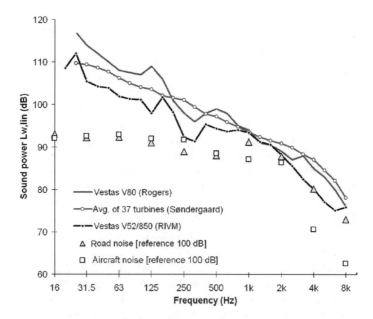

Fig. 3.12 Comparison of generalized sound spectra for different environmental acoustic sources: transportation noise via wind turbine noise

From one point of view, the existing exposure-response curve, used in any studies for impact assessments, has to be updated. From another point of view, all the non-acoustical factors, influencing on annoyance, at any specific case need to be assessed and accordingly managed correctly, providing less annoyance if it should be possible.

People are driven to complain when some nuisance factor (or *stressor*) in the environment gives rise to annoyance and when this stressor reaches a threshold of tolerance. In this context the stressor is an aircraft noise, which is described by exposure metrics usually, like in Eq. (3.12-3.14). The actual situation is rather more complex. WHO indicates also that positive well-being and quality of life can be compromised by noise annoyance and sleep disturbance first of all. Both of them are estimated for grounding the noise zoning and land use planning around the airports, using critical limits, protection guides and threshold values for sleep disturbance and annoyance [19] to control the aircraft noise impact in usual way.

The extent of noise annoyance is clearly influenced by numerous non-acoustic factors such as personal, attitudinal and situational in addition to the amount of noise per se [36]. Different models have been developed that aim to provide insight into the processes that result in noise annoyance [25, 37, 38]. However, all these models are developed based on empirical evidence related to previously found results of correlation analysis or multiple regression analysis between noise annoyance and other variables. Both these methods have severe deficiencies in modelling noise annoyance; even the direction of causation may remain uncertain. The results of correlation analysis can be misinterpreted since the effect of the factor under investigation is not controlled for noise exposure or other factors [39]. Also, in [38] it was noted that "many of the models which are tested by using path analysis are exploratory. As such, they probably do not adequately represent the processes leading to the outcome in question e.g., noise annoyance".

3.4 Non-acoustical Factors of Noise Annoyance: An Inventory

Any possible relationships between environmental noise and reported annoyance, which usually may vary from one study to another, show both direct and indirect routes from stimulus to effect. Different acoustic metrics L measure the physical amount of noise (sound exposure or peak levels); they do not measure actual community response to the noise. Approximately one-third of the variation (even only 18% by some results!) in noise annoyance can be explained by acoustical factors and a second third by non-acoustical factors [25], for which individual responses to noise among the population may vary considerably, but the social context has also been found to be important [37]. In this context the main focus is current research on noise annoyance – how they are influenced by non-acoustic as well as acoustic input variables [40]. Community response against aviation noise is closely related to perception, attitudes and expectations of the population under the impact of this noise as it follows in Table 3.5. The most important determinants of any of the

Table 3.5 Acoustic and non-acoustic factors of noise annoyance

Acoustic factors	Non-acoustic factors	Variable	Influence and effect
Sound level: definitive for exposure level of noise	*Noise sensitivity*: it is widely recognized that there are individual differences in sensitivity to noise	Noise exposure L_{DEN}	**Strong** 0.02
Frequency: definitive for perceiveness contribution in noise exposure	*Fear of noise source*: interdependence with other dangerous factors for environment, like third-party risk and air pollution around the airports	Concern about negative health effects of noise and pollution	**Strong** −1.15
Duration: definitive for perceiveness contribution in noise exposure	*Personal benefits and costs of airport operations*: employment at the airport or industry concerned; compensation and home ownership can be determined objectively, but these factors can still have a psychological effect	Positive social evaluation of noise source	**Strong** −0.05 to −0.40
Number of noise events: number of flights during the period of observation	*Attitude towards noise source authorities*: awareness of any benefits – economic and social – of the commitment generating the noise such as an airport, awareness of the costs for noise control program, fear of aircraft crashes, etc.	Negative attitude towards noise source authorities and the noise policy	**Strong** 0.11 … −0.22
Spectral composition: definitive for perceiveness contribution in noise exposure, including impact of tonal components in spectrum, impulse noise (with specific frequency component), etc.	*Perceived predictability*: this deals with the perceived probability of forecasting sound level increases from aircraft noise	Perceived disturbance	**Strong** 0.56
Amplitude fluctuation: regular and/or irregular amplitude fluctuations, more appropriate to railway noise	*Awareness of non-noise source problems*: interdependence with other dangerous factors like third-party risk and air pollution around the airports	Concern about negative health effects of noise and pollution	**Strong** 0.59
Seasonal and meteorological conditions: long-term annoyance is slightly but statistically significantly, higher in the summer than in the winter; differences exist in climate at different locations	*Perceived control and coping*: either by the individual in vicinity of the airport or by airport authority	Perceived control and coping capacity	**Strong** −0.51

(continued)

Table 3.5 (continued)

Acoustic factors	Non-acoustic factors	Variable	Influence and effect
Level of background noise: complementary to change of noise environment	*Expectations and predictability*: this deal with perceived probability of current sound level of aircraft noise, which is important for case to be protected – to be less vulnerable but also whether the information used to come to the decision is accurate and relevant	Negative expectations related to noise development	**Intermediate** 0.26 to −0.42
Flight route: predictor for noise event	*Preventability, trust and recognition:* the latter deals with recognition that impacts are recognized by local and airport authorities, and whether the information provided about the process and the decision is clear and appropriate, and whether authorities are free from bias and whether people trust their motives		**Strong**
Change in noise environment: complementary to level of background noise	*Noise insulation/compensation:* noise insulation or/and residence cost-covering schemes	Belief that noise can be prevented	**Strong** 0.03 to 0.24
	Level of background noise: rural locations with low background noise are supposed to be more vulnerable conditions for aircraft noise events		**Intermediate**
	Voice: this deals with the extent to which people are able to talk to and be heard by authorities, or in other words whether there are opportunities to participate in the decision-making process or at least whether the opinions of all parties involved are taken into account in noise management		**Strong**
	Accessibility to information: this could be seen as a means of influencing general attitudes, but it can also affect the extent to which authorities are perceived to be taking an interest in the noise-exposed community		**Intermediate**
	Home ownership: home owners might be concerned about effects on the value of their property	Concern about property devaluation	**Intermediate** 0.08 to−0.15

factors (variable) are shown also, so as their effect on noise annoyance as well (the data for standardized total effects of each variable are taken mostly from [45]). Vader in [48] compiled an array of 31 non-acoustic factors and classified each of them according to two dimensions: its *influence* on annoyance (strong, intermediate, weak – also shown in Table 3.5) and the possibility to be modified.

The general model of noise annoyance, useful for further understanding of assessment and management of the risk of human annoyance from aircraft noise, can be summarized as follows [42, 43]:

1. Population in number N (look in Eq. (3.8)) (so as any of its individuals are) is subjected to a certain noise exposure L, preferably defined as noise index L_{DN} or L_{DEN}, which is usually very accurately determined in vicinity of the airport.
2. This external determinant – noise exposure L – leads to an internal (and usually individual also) psychophysical stimulus S, the perceived noise dose, estimated with loudness (e.g. due to Stevens' type of the human ear response, technically realized as filter of type "A" [23]) scale. Stimulus S should be averaged for total population under consideration subjected to the same noise exposure L.
3. Individuals differ in their response r to the stimulus S (and accordingly to noise exposure L). The exponential distribution is chosen [42, 43] to approximate best the various possible individual reactions r to stimulus S.
4. If reaction r is more intense than a certain criterion level A, then the individual reports high annoyance (HA) – a contribution to *individual risk* (R_i) (look in Eq. (3.7)) to be annoyed by noise may appear and be identified. Internal process shown in Fig. 3.13 (from [43]) is usually influenced by external vulnerability factors, so reaction may be expected inside a specific vulnerability corridor as it was described in a Chap. 3.3 before. So, external processes in Fig. 3.13 must include not only the noise source exposure, but the vulnerability effects also.
5. Individual reaction (perception) may be changed (stressing conditions reduced), if his/her attitude and expectation to the source of noise will be changed, for example, by engagement of any individual to the process of noise management in area of concern.Thus, not only noise (stressor) management is expected to be used for receptor reaction control, but the receptor vulnerability management also.

Concluding this model one may recognize the noise annoyance as a form of psychological stress, that is, determined by the extent to which a person perceives a threat [43], i.e. perceived disturbance, and the possibilities or resources that a person has with which to face this threat. This conclusion is possible to be considered as fundamental for risk assessment and management methodology, and it is proposed to be used for noise (or particularly aircraft noise) impact assessment and management. The methodology provides necessary tools to include in consideration *vulnerability* and *capacity* values, both very important for management of the impact first of all.

Noise annoyance is defined as a form of psychological stress [42], which is determined by the perceived impact of a stressor and the perceived resources to cope with this stressor. From the point of view of psychological stress theory, the

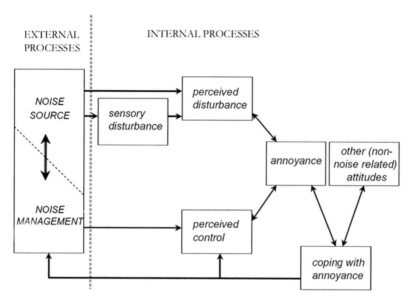

Fig. 3.13 Noise annoyance modelled as a stress response to the external stimuli "noise" and "noise management" [43]

generation of noise-induced annoyance is essentially a dynamic process. In [25] it is also emphasized that noise annoyance is not just reflecting acoustic characteristics: "noise annoyance describes a situation between an acoustic situation and a person who is forced by noise to do things he or she does not want to do, who cognitively and emotionally evaluates this situation and feels partly helpless", so how much a receptor (human) is vulnerable to stressor (noise).

Acoustical and non-acoustical factors are valued and revalued by the humans on the basis of their needs and the resources available to satisfy them. Here a primary appraisal is the level of perceived disturbance, which is evaluated by any person for the impact of the threat or harm in relation to their well-being. After a threat or harm is recognized, a process of secondary appraisal is triggered. The nature of annoyance is rooted in the fact that the exposure to noise makes it difficult or impossible to attain something valued, that is, the nature of disturbance, and two factors determine the deepness of the disturbance [43].

The structure in Fig. 3.13 represents the popular noise annoyance model: non-acoustical factors are considered as influencing the interconnection noise annoyance with sound exposure of this noise [43]. Humans are provided with the ability to create internally appropriate or at least nonconflicting attitudes towards their environment. Next to the noise management by the source, other non-noise-related attitudes can be considered as potential coping resources. Based on the model [43], it was argued that noise annoyance will arise if the perceived threat, i.e. noise, is larger than the perceived resources to face the threat, i.e. perceived control and coping capacity. In addition, even though the perceived disturbance may be very high, no noise annoyance will arise if there are sufficient coping resources. And in fact such strategies can have both positive and negative outcomes. The effects of noise annoyance on perceived disturbance and perceived control and coping capacity are equal to 0.90 and 0.94, respectively. It can be concluded that to a large extent, the reciprocal effects between noise annoyance and perceived disturbance and noise annoyance and perceived control and coping capacity cancel each other out.

The protection of the population from noise impact is understood as a dynamic process, meaning that the evaluation criteria must be repeatedly tested and – if necessary – adapted to new scientific findings [44]. The only significant determinant of perceived disturbance is the level of noise exposure. Thus through the effective management and control of aircraft noise, best practice – through ICAO Balanced Approach, it must be possible to minimize adverse impacts of aircraft noise on health and quality of life. The latter identifies a novel approach to noise mitigation using non-acoustical measures, taking advantage of those factors that have quite an impact on community response, while being modifiable by airport authorities.

Not only sounds produced are the external stimuli to which one responds, the management of sound production is an equally significant external determinant of annoyance. Until now interest in noise prevention beliefs has been moderate and even absent in some recent studies of unsteady states. New communication technologies must provide better understanding of the problem to the community, to every individual living around the airports, providing their more positive response to aircraft operation and noise in consequence.

3.5   Acoustic Modelling and Monitoring

Acoustic modelling around airports currently is intended to satisfy the needs of many users and ranges between sophisticated noise spectrum modelling and noise environment assessment in terms of cumulative noise exposure or even, by means of dose-response relationships, in terms of the size of the noise-annoyed population in the area of concern. It must be noted that the form and structure of noise indices, which we must assess and investigate around the airport or under the particular flight path, have the predominant role on the method we have to use for their assessment. Methods for modelling aircraft noise (sound radiation, propagation and attenuation) include both analytical and semiempirical results. The current tendency is towards less empirical and more analytical techniques. In general prediction schemes for aircraft noise calculation are based on three basic components:

- *Emission model:* noise radiation model, it defines sound pressure level L_{pe}, usually at reference distance 1m, with correction on directional effect (directivity pattern) K_Q and A-weighted, for the description of the emission by noise source under consideration;
- *Propagation model:* model of sound propagation/transmission from source to point of noise control, it defines the attenuation terms ΣAi due to their (sound

attenuation and refraction in atmosphere, sound diffraction around obstacles along propagation path, sound wave divergence, sound transmission through obstacles along propagation path) influences on the propagation path;
- *Immission model:* noise impact model at the control point, it defines correction value K_{IO} to describe the influences of the choice of the immission point on sound levels detected by receptor.

Noise levels such as L_{Aeq} (L_{DN}, L_{DEN}, *WECPNL*, etc.) provide a good indication of the amount of physical noise present at point of its perception, and changes in physical noise level can be expected to correlate with changes in resident's perception, at least to some degree. In determining the noise level at point of consideration, a model should be used as a basis that describes the calculated immission level as a function of *emission, propagation* and *immission models*:

$$L_{pi} = L_{pe} + K_Q + \Sigma A_i + K_{IO} \qquad (3.15)$$

where L_{pi} is A-weighted sound pressure level at the place of *immission*; L_{pe}. The uncertainty of the sound level values determined by Eq. (3.15) is between 3 and 5 dBA, depending on type of traffic noise in general, and it is equal for aircraft noise.

The *immission model*, as a component of the overall noise modelling scheme, is a subject of current deeper consideration, with psychological phenomenon of annoyance inside and risk (hazard, vulnerability and coping capacity) methodology as a tool to assess. Percentage of highly annoyed (*HA*) people in dependence from day-night average sound level L_{DN} for three modes of transportation is a classical picture of *population vulnerability to the type of transportation noise source* (Fig. 3.6b) – it is the highest for aviation in comparison with road and railway noise effects. For the level $L_{DN} = 65$ dBA (a limit proposed to be used for prohibition of residential area in airports vicinity), the percentage of *HA* people under the aircraft noise is expected to be 30%, while for railway noise it is equal to 12% only, so the noise exposure level is the same, but damage for population (their annoyance) is assessed as two to three times higher. For this reason, the standard ISO 1996-1:2003 [23] recommends 3–6 dB penalties and bonuses (or simply adjustments) for aircraft and train noise, respectively. Moreover, the response curves were changed dramatically during the last decades (Fig. 3.5), becoming more "annoying" in comparison with their first definitions [21], it may be concluded that exposure levels are still the same, never mind that flight traffic grew sufficiently, but vulnerability effects are changed huge (community expectations first of all).

In an attempt to reduce the scatter to the community response data (Fig. 3.11), the EPA [21] suggested the use of "normalized" L_{DN}, which is the measured or predicted L_{DN} with a number of *adjustments* like in Eq. (3.11) added to account for specific characteristics of the sound (Table 3.6 shows the EPA-suggested adjustment factors and their magnitudes [46]). All of them in proposed above risk terminology are *vulnerability factors* for the risk to be annoyed by noise assessment also. For new situations, especially when the community is not familiar with the sound source in question, greater community annoyance than predicted by application of the

Table 3.6 Vulnerability effect adjustments (EPA-recommended adjustments in [21]) to be added to the measured or predicted L_{DN} of an intruding noise at a residential location [46]

Type of adjustment	Description of condition	Adjustment to be added to measured L_{DN}, dBA
Seasonal considerations	Summer (or year-round operation)	0
	Winter only (or windows always closed)	−5
Adjustment for *outdoor background noise* measured in the absence of intruding noise (*change in noise environment*)	Quiet suburban or rural community (remote from large cities and from industrial activity and trucking)	+10
	Normal suburban community (not located near an industrial activity)	+5
	Urban residential community (not immediately adjacent to heavily travelled roads or industrial areas)	0
	Noisy urban residential community (near relatively busy roads or industrial areas)	−5
	Very noisy urban residential community	−10
Adjustment for previous exposure (*change in noise environment*) and *community attitudes*	The community has no prior experience with the intruding noise	+5
	Community has had some previous exposure to the intruding noise, but little effort is being made to control the noise. This adjustment may also be applied in a situation where the community has not been exposed to the noise previously, but the people are aware that bona fide efforts are being made to control the noise	0
	Community has had considerable previous exposure to the intruding noise, and the noisemaker's relations with the community are good	−5
	Community is aware that the operation causing the noise is very necessary and will not continue indefinitely. This adjustment can be applied for an operation of limited duration and under emergency circumstances	−10
Pure tone or impulsive sound	No pure tone or impulsive character	0
	Pure tone or impulsive character present	+5

equation can be expected ("expectations and predictability" – non-acoustical factor in Table 3.5); the difference may be as much as +5 dB. One more classical example of noise impact vulnerability is additional guideline values, which are suggested for specific environments [7] shown in Table 3.7, all data are in L_{Aeq}.

Looking in Eq. (3.10) and considering the noise annoyance effect, it was proposed to represent the likelihood $P_{d/f}$ as a dependence of $HA\%$ from noise metric L_{DN} (or its analogue L_{DEN}), currently it should be noted that normalized dependence is considered. A vulnerability shift in relation to noise source (ΔL_s) is proposed to be

Table 3.7 WHO noise guidelines, 1996 [7]

Daytime		Night-time		
Inside	Outside	Inside	Outside	Type of residence
50 dBA	55 dBA			Dwellings
		30 dBA	45 dBA 45 dB L_{Amax}	Bedrooms
35 dBA	55 dBA			Schools
35 dBA 30 dBA		35 dBA 45 dBAmax 30 dBA 40 dBAmax		Hospitals *General* *ward rooms*
100 dBA for 4 hours period				Concert halls

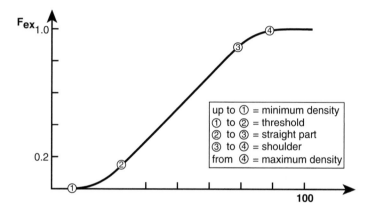

Fig. 3.14 Factor of expectation (expectation factor $F_{ex} = [0,1]$) in dependence with rate of expectation $R_{ex} = [0,100]$: any deviation from the expected level in the direction of growth causes the growth F_{ex}

included in a form of adjustment used in [23] – Eq. (3.11). Today it is highest for noise from wind turbines (Fig. 3.5), because expectation rate among the population in quiet suburban or rural community, where wind farms are usually installed, is highest. Such expectation rate is introduced to assess the expected vulnerability effect on a value of response of the population on noise via the factor of expectation (Fig. 3.14):

$$\Delta L_{s\,i} = \Delta L_{s\,i\,max} F_{ex},\qquad(3.16)$$

where i is a type of vulnerability considered and $\Delta L_{s\,i\,max}$ is a maximum possible value of vulnerability shift.

Further step is a "normalization" procedure for noise level used in noise impact assessment:

$$L_{DN\,norm} = L_{DN\,cal/meas} + \Delta L_{s\,\Sigma}$$

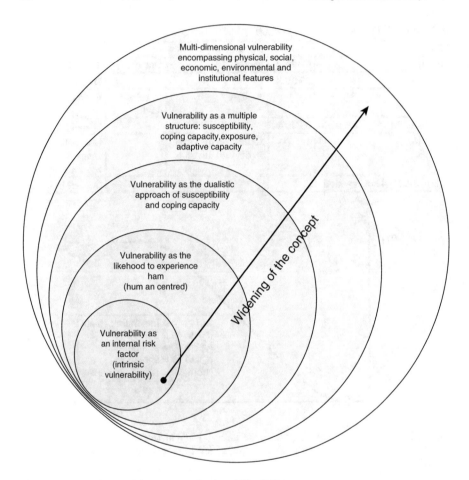

Fig. 3.15 Key spheres of the concept of vulnerability [12]

where calculated or measured value $L_{DN\,cal/meas}$ is correspondent with case of noise event under consideration, and vulnerability shift $\Delta L_{s\Sigma}$ may include additively a number of factors influencing on vulnerability of the receptor to noise in this case. Concept of vulnerability is proposed to be widening to include the coping capacity of the system under consideration, as it is considered by [12], Fig. 3.15, and it takes into account the multifunctional dependence between hazard, vulnerability and capacity (concept-formula Eq. (3.4)). In such context, in dependence with conditions and requirements for aircraft noise (or environmental noise in general) impact assessment in values of human annoyance to noise the main terms in Fig. 3.15 are defined as following: the *intrinsic vulnerability* is a known receptor (human) sensitivity to noise; *human centred experience with ham* (noise annoyance) is a differentiation between the receptor susceptibility to source of noise and other possible vulnerability factors (Tables 3.1, 3.6 and 3.7); *dualistic susceptibility/coping capacity* approach defines the changed susceptibility of the receptor to noise impact if any kind of protection measure is realised at a receptor point at the moment of noise impact assessment; *multiple structure of the vulnerability* to noise annoyance is a balanced approach to noise exposure control (in source, along propagation path, at receptor point); *multi-dimensional vulnerability* is combined exposure control with control of social, economic, environmental and institutional factors (Table 3.1) of vulnerability.

For example, in considering the effects of the number of aircraft movements on annoyance, we must acknowledge that term "number of events" N_{events} is already included in the L_{Aeq} type (or L_{DN} or L_{DEN}) of noise level descriptions. The common energy equivalent continuous sound level assumes that each doubling or halving of the numbers of noise events is equivalent to a 3 dBA increase or decrease in average sound levels ($k = 10$ in component $k*lg(N_{events})$ of the calculation equation for L_{Aeq}). The ANASE study [15] concluded that a k-factor of >20 would be more appropriate, because "people are more sensitive to increased aircraft numbers than assumed by L_{Aeq}". The results of ANASE suggest that there would be no trend in aircraft noise annoyance over time in case the number of movements would be given more weight in the L_{Aeq} formula. The recent study [47] compared 32–39 aircraft noise surveys (1973–2015) with respect to the total number of flights at the surveyed airports and concluded that CTL values (i.e. the L_{DN} value at which 50% of the respondents are highly annoyed) decreased with increasing numbers of flight movements. In other words, especially at large airports with high numbers of aircraft movements, more survey participants were highly annoyed at lower L_{DN} values.

Besides calculation module the monitoring system involves a subsystem of continued observation, measurement, forecasting and evaluation for defined purposes and is the basic tool for that underpins responsible environmental management [5]. Noise monitoring to be undertaken usually in their local community on the assumption that aircraft noise will exceed what is considered "acceptable" or legally permissible, and in this connection it is necessary to refer to the legislative controls on aircraft noise. In general case the purposes of monitoring are described elsewhere as (1) to assess the current status of the resource to be managed or to help determine the priorities for management, (2) to determine if the desired management strategies were followed and produced the desired consequences and (3) to provide a greater understanding of the system being managed. The number and location of the monitors are important depending upon the specific role they are to play. Quite usual elements of current aircraft noise monitoring system are the air traffic data connection for flight events detection (correlation with noise events) and gathering the complaints from residents living around - to detect how the receptors of noise realy react to this noise in their environs. Such systems can be used to [5]:

- Compile data on methods used to describe aircraft noise exposure.
- Determine the contribution (general and/or specific by type, route, airlines, etc.) of aircraft to the overall noise exposure.
- Collect data on the characteristics of airports with noise and/or flight path monitoring systems.
- Research and develop better noise abatement procedures specific to the airport.

- Collect details of airport noise monitoring systems such as capabilities, data stored and technical support.
- Compare calculated and monitored noise levels for a suitable sample of airports.
- Compare measured noise levels with certified noise levels for a range of aircraft types and operating conditions.
- Measure the effectiveness of a noise control programme.
- Examine changes in measured noise exposure over a representative time period.
- Allow the correlation of complaints with noise events.
- Update advisory documents on methodologies and applications of noise contouring and monitoring, supplemented, for environmental noise management, by the elements of expert and decision-making systems.

This collection of information:

- Enables assessment of the effects of operational and administrative procedures for noise control and compliance with these procedures and/or assess alternative flight procedures for noise control (the tool of objective assessment of efficiency of the proposed operational and administrative procedures for noise control in the vicinity of the airport).
- Assists in the planning of airspace usage issues.
- Increases public confidence that airport-related noise is being monitored to protect the public interest.
- Enables validation of noise forecasts and forecasting techniques and their methodologies over an extended period of time (collection of data for noise contouring, system noise exposure forecasting and contouring with compiled data).
- Assists relevant authorities in land use planning for developments and noise impact on areas in the vicinity of an airport.
- Enables assessment of a Quota Count system (special mitigation procedure which defines a number of flights of the aircraft of specific types during a specific period of the day), among other possible noise mitigation measures.
- Indicates official concern for airport noise by its jurisdiction and its governing bodies and enables provision of reports to, and responses to, questions from government and other members of parliament, industry organizations, airport owners, community groups and individuals.

Not only sounds produced are the external stimuli to which receptor responds, the management of sound production is an equally significant external determinant of annoyance. Until now interest in noise prevention beliefs has been moderate and even absent in some recent studies. New communication technologies must provide better understanding of the problem to the community, to every individual living around the airports, providing their more positive response to aircraft operation and noise in consequence.

Currently the common characteristics of the airports to manage the issue of concern are the following:

- Limited technical resources available (skills, tools, infrastructures) for noise prevention and/or control;

– Limited financial resources available, especially for noise prevention and/or control;
– Relatively low traffic volumes in prevalent number of airports;
– Wide range of aircraft classes in operation, with single noisy events (dominant for total noise exposure) for heavy/noisy aircraft/helicopters;
– Relatively low integrated noise indicators with relatively high-noise events, often between 10 and 20 during night period

3.6   Balanced Approach for Aircraft Noise Impact Reduction

Current ICAO [8] and ACARE [41] targets and goals are not only to reduce the noise levels, *the novel and more real approach is based on the idea that noise level reduction at receiver point is not a final result for society, but it is just a tool to achieve the real final goal, which is the reduction of the noise effects.* By ICAO this effect is defined currently as a reduction of number of people affected by aircraft noise – or simply a number of exposed people by noise over the protection guide value or predefined *number of highly annoyed people*. This is the reason for a number of new current concepts, approaches and efforts to reduce aviation noise annoyance (sleep disturbance and other effects of noise impact), keeping the produced noise levels the same (Fig. 3.16).

Protecting residents is a dynamic process that must be followed up. The evaluation limits must be repeatedly tested in view of new scientific findings and adapted, if necessary. *Traditional* taken *approaches to aircraft noise management* include reducing aircraft noise at source, to devise operational procedures and restrictions, flight routes, and other forms of mitigation, etc., to minimize individual residential exposure via ICAO balanced approach (BA) to aircraft noise control around the airports [49] and to keep the public fully informed about noise management and noise control [50, 51]. The main objective is that noise problems can be addressed in an environmentally and economically responsible manner within the system while preserving potential benefits gained from aircraft-related measures.

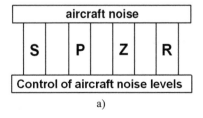
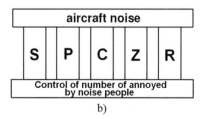

a)   b)

Fig. 3.16 Two approaches for aircraft noise impact management in airports: (**a**) 4-pillar balanced approach by ICAO (**S** – in source; **P** – noise abatement procedures; **Z** – noise zoning and land use; **R** – flight restrictions); (**b**) new approach including communication (**C**) with population to control its reaction to noise

Fig. 3.17 An algorithm
for noise management

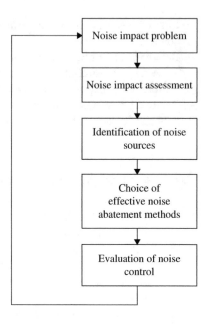

Framework for risk assessment and reduction, shown in Fig. 3.7, is correct for aircraft noise impact management, the Fig. 3.17 shows the stages for noise management in more simple form, in general it is correct for BA approach implementation completely, so as it is correct for any specific its element. Noise impact assessment, including the terms of risk to be annoyed by noise, must be finished with comparison of assessed (modelled or monitored) sound levels of noise with installed limits by the national and international standards. If the assessed levels are over the limits a further analysis should be done to define the sources with dominant contribution to noise exposure at point (area) of noise control. Id such a source is an aircraft flight event (or airport) – the ICAO BA should be realized as mostly effective approach nowadays, better in combination with continuous monitoring of the sound levels (noise indices), etc.

Tremendous results were achieved during the last decades in any of BA element implementations in a world. Especially the huge noise reduction has been reached in a source, which was outlined further in new more stringent standards from ICAO (Fig. 3.18) and in national ones. Thanks to technology, today's aircraft is 50% quieter than 10 years ago. Research initiatives target a further reduction of 50% by 2020. Any new steps in noise generation reduction are currently much more difficult, because all the character acoustic sources of the aircraft are balanced between themselves (Fig. 3.19), so if the task of following overall aircraft noise reduction will be formulated, its solution should be found for all the sources at the same time, once again in balanced way – as acoustically, so as economically.

Flight procedures for noise reduction in flight operation are much wider today, even when they are recommended by ICAO circulars [52]. Operational opportunities for reducing aircraft noise were identified across a variety of flight phases, covers ground, departure and arrival operations, for aeroplanes and helicopters, ATM

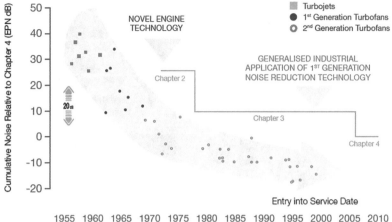

Fig. 3.18 Cumulative stringency comparison of ICAO standards for aircraft noise

contribution, etc. It is recognized that noise is a complex subject and that operational procedures designed to minimize the sound levels of aircraft operations at one point along the flight path may result in increases in noise impacts elsewhere. In this situation, it is important to understand what the noise impacts are along the whole of the flight path of the single flight event and around the area of control in vicinity of the airport.

A number of noise abatement procedures (NAP) involve a power reduction at or above the prescribed minimum altitude (distance to point of noise control) and delaying flap/slat retraction until the prescribed maximum altitude is attained. But it should be noted that if a more stringent noise standard implemented as a result of more balanced acoustic sources contribution in a reduced overall aircraft noise – Fig. 3.19, NAP will be less efficient for any type of the aircraft, because the less aircraft_noise/engine_mode gradient provided by quieter aircraft design (Fig. 3.20).

Till now all the existing BA elements are subjects to identify and assess the noise exposure, mostly via noise contour modelling, in some cases via monitoring, which allows to evaluate noise control measures and to determine the most cost-effective and efficient for environment protection set of them [49]. In best known solutions, the process is continuing with public notification and consultation procedures and even being a mechanism for dispute resolution. This important approach is implemented in the European Environmental Noise Directive [53]. According to it, noise action plans will be developed with the participation of the public. The claim of the citizens in participation has steadily grown, especially if their residential area or essential environmental aspects are concerned.

With this context of public involvement it is appropriate to begin with new vision on ICAO BA to aircraft noise control (namely, to add to the existing elements of noise reduction: at source, by noise zoning and land use planning, with operational procedure and mitigation measures) the newish element – the *reduction of the noise effects* via *novel concept*, approaches and efforts *to reduce aviation noise annoyance,*

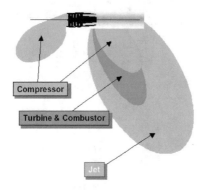

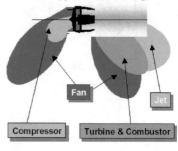

Noise of a typical 1960s engine

Noise of a typical 1990s engine

Turbojets dominated by high jet exhaust noise at departure - loud roar, rumble

High-bypass-ratio turbofans dominated by fan noise - whine, whistle - and lower jet exhaust noise - roar, rumble

a)   **b)**

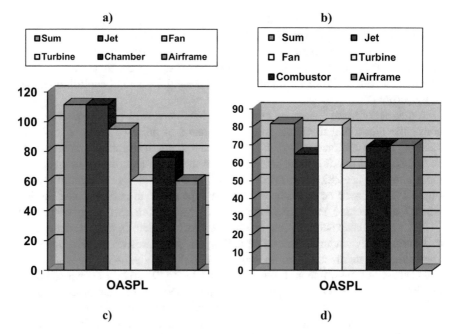

c)   **d)**

Fig. 3.19 Overall noise performances for aircraft with bypass engines: (**a**) noise directivity patterns for engine with bypass ratio $m \sim 1$; (**b**) noise directivity patterns for engine with bypass ratio $m \sim 6$; (**c**) acoustic source contribution to OASPL of the aircraft with engine $m \sim 1$; (**d**) acoustic source contribution to OASPL of the aircraft with engine $m \sim 6$

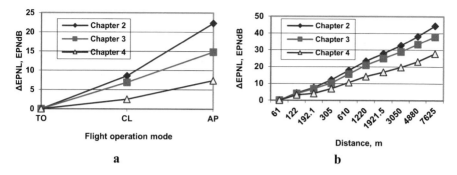

Fig. 3.20 Aircraft noise standard stringency influence on NAP ability to reduce noise level: (**a**) at point of noise control; (**b**) along flight path distance

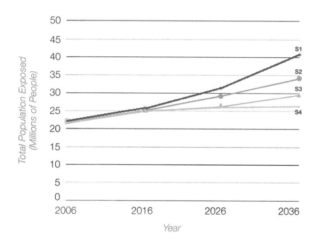

Fig. 3.21 Total global population exposed to aircraft noise above 55 DNL: CAEP forecasting for noise reduction in preparation to new ICAO standard (Chap. 14 of Annex 16, vol. 1) for aircraft noise control [54]

or more directly – to reduce vulnerability effects for the receptors effected by noise, and accordingly a number of affected people by noise.

It is important to differentiate between noise exposure and the resulting noise nuisance in different communities and manage each appropriately. The type of information collected and the way in which it is analysed and reported will differ according to the objective of the programme of noise control. Usual option of quantifying overall noise exposure – through noise contour modelling and quantifying the number of people inside the contour with specified noise level (predefined by rules for noise zoning around airports or by general noise control rules like Environmental Noise Directive or CAEP documents, see Fig. 3.21).

This is the reason for a number of current concept, approaches and efforts to reduce aviation noise annoyance, keeping the produced noise levels the same. This objective is expected to be achieved by bringing information closer to the people living in airport surroundings. For example, there is some previous assumption that shared unattended noise monitoring results can improve airport noise acceptance, as general public can check the compliance with noise limits in their proximity, raising people awareness [5].

Comparing with traditional BA elements, which are defined by physical effects of sound generation and propagation, an annoyance is a psychological phenomenon (in nature of effect on humans, the noise is a psychological phenomenon too!). Acoustical factors of environment noise events like sound intensity, peak levels, duration of time in-between sound events, number of events, etc., were focused for explanation of annoyance mainly [55]. The non-acoustical factors ("moderators" and/or "modifiers" of the effect) have still received an empirical attention, without deep theoretical approach, never mind that various comparative studies reveal that they play a major role in defining the impact on people [37].

To evaluate the effect of the protection measures implemented, a comprehensive set of surveys to evaluate the short- and long-term effects should be undertaken. A number of the previous studies indicate that when changes in noise exposure are achieved by source-related measures (quieter aircraft and/or low noise flight proce-dures implemented, air traffic reduced, etc.), the responses could be higher than those predicted from the exposure-response relationships established from a more stable condition. In studies where the changes include noise screens or insulation efforts, the change may be smaller than predicted. For example, inside dwellings of the "experimental" group that received the noise reduction intervention, an average equivalent noise reduction of 7 dBA was calculated inside the dwellings. But some of intervention studies show that people are often satisfied with an intervention regardless of the result of the intervention (Hawthorne effect). For example, one study shows the positive effect equal to average equivalent noise reduction of 5 dBA from informing a population about simply a noise monitoring programme realized carefully around the airport. A review of different theoretical approaches explaining such differences can be found elsewhere.

These characteristics with the features of the problem defined before provide the necessity to solve the total problem of the aircraft noise management in and around the airports under consideration grounding on complex system approach including the following elements (Table 3.8). Such a system should continue to search for the missing components and their further inclusion inside the separate tools, taking in mind the existing and newly defined non-acoustical factors of noise annoyance at any specific airport under consideration and necessity to provide perceived control and coping capacity for noise annoyance management. For example, measurement toolbox should include two or three alternative solutions for autonomous localiza-tion of the noise events:

- Passive acoustic radar to define flight path of aircraft noise event;
- Combination of acoustic signal analysis and other sensors;

Table 3.8 Working packages (WP) of the overall toolbox for aircraft noise impact management in airport

WP	Tasks to be solved
Measurement toolbox	Direct measurements of the noise levels (exposure)
	Develop the missing components in the measurement chain
Modelling toolbox	Direct modelling of the noise exposure and people exposed to noise
	Develop the missing components in the modelling chain
Community toolbox	Direct and indirect communication with the community
	Develop the missing components in the communication with community chain
Environmental management	Integrate the toolboxes for aircraft noise in the overall environmental management toolkit
	Assess current results and provide guidelines for management
Project management	Project management in airport and community around the airport

- Aircraft identification by image detection;
- Use additional info, if available (flight plans, modes data, etc.).

Methodology for modelling toolbox includes similar approach for general aviation and scheduled air traffic using noise-power-distance concept for both [5] and provides the tools for thrust estimator and flight track handler to simplify the ability to control the noise event with time not only for noise levels but for aircraft coordinates also.

Development of community toolbox includes the elements like virtual microphone, visualization system and virtual complaint helpdesk.

Environmental management toolkit provides the interdependency analysis and assessment to cover the dependence of noise tasks and solutions from other environmental issues like air pollution and third-party risk assessment (aircraft crashes).

Project management provides overall project coordination and communication; single, joint and interdependency analysis; assessment and decision-making; monitoring, modelling and communication toolboxes on a single platform (web-based Project Data Exchange System, PDES); their organizational and financial implementation with solving of any technical, financial or administrative issues; or conflicts between the elements, when it is needed.

In best known solutions, the process is continuing with public notification and consultation procedures and even being a mechanism for dispute resolution. The type of information collected and the way in which it is analysed and reported will differ according to the objective of the programme of noise control. This objective is expected to be achieved by bringing information closer to the people living in airport surroundings - customizing it to target meaningful and friendliness, in order to optimize the aircraft noise mitigation strategies. For example, there is some previous assumption that shared unattended noise monitoring results can improve airport noise acceptance, as general public can check the compliance with noise limits in their proximity, raising people awareness. The generalized use of the Internet in recent years has allowed improving data accessibility by the general public, but (a) information reported is overly technical and should be customized for different users' profiles, so that they can understand the information provided, and (b) commonly used noise prediction indices do not satisfy the general public's expectations, as, on some occasions, they seem to mask the real pollution under mathematical operations. New studies should be implemented to test traditional and novel reporting templates aimed at improving awareness, comprehensibility and properly matching noise scenarios to people's perception.

The measurements of aircraft noise and the analysis of the results are necessary in order to protect correctly the local community living in the airport surrounding areas. Permanent or/and temporary noise monitoring to be undertaken usually in their local community on the assumption that aircraft noise will exceed what is considered "acceptable" or legally permissible, and in this connection it is necessary to refer to the legislative controls on aircraft noise. The results show that for airports with low intensity of flights, the long-term equivalent sound level is heavily changing in relation with the long-term maximum sound level, but for high intensity flight traffic, this interrelation is quite stable. In the vicinity of airports with low flight intensity, the maximum sound level as a noise impact metric is more sensitive than the equivalent level. In general case the purposes of monitoring are described elsewhere as (1) to assess the current status of the resource to be managed or to help determine the priorities for management, (2) to determine if the desired management strategies were followed and produced the desired consequences, (3) to provide a greater understanding of the system being managed and (4) to show that population involvement in noise management helps to reach the goals of the noise management program, etc. Although today in most cases the main concern is the negative impact of aircraft noise, the highest goal is to show that measuring and monitoring the aircraft noise can be used for positive purposes. For example, to show in routine mode what an aircraft exceeded the permissible level at a point of noise control, to show even why it was exceeded (flight procedure mistake happened or an aircraft type is quite noisy to be operated in particular conditions), any flight safety issues may be raised with monitoring system usage and at the same moment providing confidence to aviation as a whole. A very new challenge should be expected: how to deliver respite from aircraft noise at the airport that is valued by the community, which is consistent with efficient operations?

3.7   Conclusion

It should be a primary objective of future research into environmental noise impact to investigate the interplay of sound level control and perceived control. New and additional (political) measures to mitigate noise impact may result from the redirection of attention from sound to noise and to noise annoyance. Strategies that reduce noise annoyance, as opposed to noise, may be more effective in terms of protecting public health from the adverse impacts of noise and its interdependency with other environmental, operational, economic and organizational issues of airport and airlines operation and maintenance.

The reviewed and proposed models provide a good model fit and support to the toolboxes of noise annoyance management, currently under the design. It can be concluded that the concern about the negative health effects of noise and pollution, other environmental issues, are still the subjects of scientific and societal attention, their newish deliverables may improve the approach to build the fifth element of ICAO balanced approach to aircraft noise control around the airports, which cover the measures to reach the final goal of aircraft noise management – to reduce the number of people living in vicinity of the airports and affected by noise.

The current study can be beneficial for the researchers interested in aerospace and environmental sciences.

References

1. Last J (1987) Public Health and Human Ecology. Appleton and Lange, East Norwalk
2. Weltgesundheitsorganisation. Regionalbüro für Europa (2018) Environmental noise guidelines for European region. World Health Organization Regional Office for Europe, Copenhagen
3. Zaporozhets O, et al (2014) Interdependency analysis of current and future criteria and their limits for aircraft noise control in urban areas. Acoustic climate inside and outside buildings, international conference, 23–26 Sept 2014, Vilnius, Lithuania, abstract number: Acoustic 14
4. Kazhan K, Tokarev V (2009) The optimal noise modelling in the airport vicinity. 9th international conference on theoretical and computational acoustics, 7–9 Sept 2009 proceedings, Dresden. pp 41–50
5. Zaporozhets O, Tokarev V, Attenborough K (2011) Aircraft noise: assessment, prediction and control. Glyph International, Taylor & Francis; CRC Press, 2017, ISBN 9781138073029, CAT# K33865
6. Noisy areas may reduce life expectancy, study shows (2015) Chicago Tribune, July 15 2015. http://www.chicagotribune.com/lifestyles/health/sc-hlth-701-noisy-neighborhoods-life-expectancy-20150625-story.html
7. Berglund B et al (1999) Guidelines for community noise. World Health Organization, Geneva, http://www.who.int/iris/handle/10665/66217
8. ICAO Resolution A39-1 (2016) Consolidated statement of continuing ICAO policies and practices related to environmental protection – general provisions, noise and local air quality ICAO resolutions adopted by the assembly, Provisional edition, Oct 2016, 138 p
9. UNISDR (2016) Exposure and vulnerability short concept note: work stream 2, working group 2. UNISDR science and technology conference on the implementation of the Sendai Framework for Disaster Risk Reduction 2015-2030. 27–29 Jan 2016, Geneva International Conference Centre
10. United Nations International Strategy for Disaster Reduction (2009) UNISDR Terminology on Disaster Risk Reduction. UNISDR, Geneva, p 35
11. ISO 31000:2018. Risk management – guidelines. 2018-02, 18 p
12. Birkmann J (ed) (2006) Measuring vulnerability to natural hazards: towards disaster resilient societies. United Nations, University Press, Tokyo, New York, Paris, 524 pp
13. van Westen CJ, Greiving S (2017) Multi-hazard risk assessment and decision making. In: Dalezios NR (ed) Environmental hazards methodologies for risk assessment and management. IWA Publishing, London, pp 31–94
14. Zaporozhets O, Blyukher B (2016) Hazard analysis and risk assessment methodology: generalized model. In: Encyclopedia of soil science, 3rd edn. Taylor & Francis, Boca Raton, pp 1–15. https://doi.org/10.1081/E-ESS3-120053900

15. Flindell I et al (2013) Understanding UK community annoyance with aircraft noise. ANASE update study report for 2M Group, Ian Flindell & Associates
16. Zaporozhets OI, Khaidar HA (2001) Instruments and procedures for management of aviation safety provision. Visnyk of the National Aviation University, Kyiv, 6(1): 186–189
17. Morselt T, Evenhuis E (2006) Review standaardmethode schade en slachtoff ers, Rebel Group in opdracht Rijkswaterstaat, Dienst Weg en Waterbouwkunde, Rotterdam, 2006
18. Basner M (2014) Auditory and non-auditory effects of noise on health. Lancet 383(9925):1325–1332
19. Griefahn B, Scheuch K (2004) Protection goals for residents in the vicinity of civil airports. Noise Health 6(24):51–62
20. American Chemical Society (1998) Understanding risk analysis. A short guide for health, safety, and environmental policy making. American Chemical Society, Washington, DC. Internet edition, www.rff.org/rff/publications/upload/14418_1.pdf
21. EPA (1973) Public health and welfare criteria for noise. U.S. Environmental Protection Agency, Office of Noise Abatement and Control (ONAC), report EPA550/9-73-002, Washington, DC
22. Miedema HME, Vos H (1998) Exposure-response relationships for transportation noise. J Acoust Soc Am 104:3432–3445
23. ISO 1996-1 (2016) Acoustics—description, measurement and assessment of environmental sound—part 1: basic quantities and assessment procedures, International Standard ISO 1996-1. International Organization for Standardization, Geneva
24. Shaw EAG (1996) Noise environments outdoors and the effects of community noise exposure. Noise Control Eng 44(3):109
25. Guski R (1999) Personal and social variables as co-determinants of noise annoyance. Noise & Health 1(3):45–56
26. Guski R et al (2017) WHO environmental noise guidelines for the European region: a systematic review on environmental noise and annoyance. Int J Environ Res Public Health 14:1539. https://doi.org/10.3390/ijerph14121539
27. Aircraft Noise and Annoyance (2018) Recent findings CAP 1588. Civil Aviation Authority
28. Fidell S (2003) The Schultz curve 25 years later: a research perspective. J Acoust Soc Am 114(6):3007–3015
29. WHO (2009) Night noise guidelines for Europe. WHO Regional Office for Europe, Copenhagen
30. Wirth K, Brink M, Schierz C, Aircraft noise annoyance at different times of day. https://www.researchgate.net/publication/228790089
31. Schultz TJ (1979) Synthesis of social surveys on noise annoyance. J Acoust Soc Am 65:849
32. Miedema HM, Oudshoorn CG (2001) Annoyance from transportation noise: relationships with exposure metrics DNL and DENL and their confidence intervals. Environ Health Perspect 109(4):409–416
33. Gjestland T, Gelderblom FB (2017) Prevalence of noise induced annoyance and its dependency on number of aircraft movements. Acta Acustica United Acustica 103:28–33
34. ICAO white paper on noise (2018) CAEP/11-WP/xx, draft. 41 p
35. Jagnatinskas A, Fiks B, Zaporozhets O, Kartyshev O (2011) Aircraft noise assessment in the vicinity of airports with different descriptors. Inter-noise-2011, Osaka, Japan, 4–7 Sept 2011
36. Scheuch K, Griefahn B, Jansen G, Spreng M (2003) Evaluation criteria for aircraft noise. Rev Environ Health 18(3):185–201
37. Job RFS (1988) Community response to noise: a review of factors influencing the relationship between noise exposure and reaction. J Acoust Soc Am 83:991–1001
38. Taylor SM (1984) A path model of aircraft noise annoyance? J Sound Vib 96:243–260
39. Alexandre A (1976) An assessment of certain causal models used in surveys on aircraft noise annoyance. J Sound Vib 44:119–125
40. Flindell IH, Stallen PJM (1999) Non-acoustical factors in environmental noise. Noise Health 1(3):11–16
41. ACARE (2012) Realising Europe's vision for aviation. Strateg Res Innov Agenda 1, 150 p. www.acare4europe.org

42. Fidell S, Piersons KS (1997) Community response to environmental noise. In: Crocker MJ (ed) Encyclopaedia of acoustics. Wiley, New York
43. Stallen PJM (1999) A theoretical framework for environmental noise annoyance. Noise Health 1
44. Zaporozhets O (2014) Criteria for aircraft noise control around airports and their role in reaching the strategic goals in environmental protection from aviation impact. Acoustic climate inside and outside buildings, international conference, 23–26 Sept 2014, Vilnius, Lithuania, abstract number: Acoustic 09
45. Kroesen M, Molin EJE, van Wee B (2008) Testing a theory of aircraft noise annoyance: a structural equation analysis. J Acoust Soc Am 123(6): 4250–4260
46. Schomer PD (2005) Criteria for assessment of noise annoyance. Noise Control Eng J 53(4):132–144
47. Guski R (2017) The increase of aircraft noise annoyance in communities. Causes and consequences, ISBEN-2017, Keynote lecture, Zurich, 12 p
48. Vader R (2007) D/R&D 07/026 Noise annoyance mitigation at airports by non-acoustic measures
49. ICAO (2004) Guidance on the balanced approach to aircraft noise management. ICAO Doc 9829, AN/451, Montreal
50. Woodward JM, Lassman Briscoe L, Dunholter P (2009) Aircraft noise: a toolkit for managing community expectations. ACRP report 15, Washington, DC
51. ICAO Circular 351 (2016) Community engagement for aviation environmental management. ICAO Cir. 351-AT/194, 2017
52. Review of noise abatement procedure research & development and implementation results. Discussion of survey results ICAO preliminary edition – 2007, 29 p
53. Directive 2002/49/EC of the European Parliament and of the Council of 25 June 2002 relating to the assessment and management of environmental noise, OJ L 189, 18.7.2002
54. Fleming GG, Ziegler URF (2013) Environmental trends in aviation to 2050. ICAO 2013 Environmental Report
55. Janssen SA, Vos H, van Kempen EE, Breugelmans OR, Miedema HM (2011) Trends in aircraft noise annoyance: the role of study and sample characteristics. J Acoust Soc Am 129(4)

Chapter 4
Biodiversity Management

Hulya Altuntas

Biodiversity is generally defined as the variety of life on Earth. Therefore, biodiversity is the main resource, which is indispensable for providing the basic vital needs of people. In general, biodiversity is evaluated in three hierarchical categories: genetic diversity, species diversity, and ecosystem diversity. Human factor plays a role directly or indirectly at the beginning of the factors that negatively affect biodiversity in the Earth. Conservation, and sustainable management of biodiversity are a common responsibility of people. For this reason, biodiversity management has an important place in aviation as in every field. In this chapter, firstly, biodiversity, the reasons of biodiversity loss, and biodiversity conservation are discussed. Then, biodiversity management and methods are described in four stages: definition of the problem, planning, monitoring of biodiversity, and evaluation. Assessments on the applicability of these methods in aviation are also discussed in this section. Finally, the importance of sustainable management of biodiversity at airports is evaluated.

4.1 Fundamentals of Biodiversity

The expansion of the human population has caused the extinction of hundreds of species and thousands of species to face extinction danger. These changes indicate the loss of biological diversity. The loss of biological diversity is directly affected by the disappearance of natural ecosystems [1–4]. Only a quarter of the world's land has remained unaffected by the changes caused by people, and as such, evidence of the effects of human activities on natural ecosystems are increasing as time goes on [5]. For example, oil spill in the Gulf of Mexico, in 2010, caused billions of dollars losses in fishery, recreation, and other industries. At the same time, studies estimate that 800,000 birds died because of exposed BP oil [3–6]. As understood in this

H. Altuntas (✉)
Eskisehir Technical University, Faculty of Science, Department of Biology, Eskisehir, Turkey

© Springer Nature Switzerland AG 2019   81
T. H. Karakoc et al. (eds.), *Sustainable Aviation*,
https://doi.org/10.1007/978-3-030-14195-0_4

example, the importance of biological diversity in the wise management of Earth's resources is emphasized in this chapter. Biodiversity is an abbreviation of natural/ biological diversity and is the variety and variability of life on Earth (microorganisms, fungi, plants, and animals). Biodiversity also includes the genetic diversity of organisms in all habitats [5, 7, 8].

The Convention on Biological Diversity (CBD), developed and signed by 157 governments at the Earth Summit Rio de Janeiro in 1992, today having 193 member states, defines biodiversity in the following manner:

> Biological diversity means the variability among living organisms from all sources including, inter alia, terrestrial, marine and other aquatic ecosystems and the ecological complexes of which they are part; this includes diversity within species, between species and of ecosystems [8].

According to this definition, the CBD covers three equally consequential and supplementary objectives:

1. The preservation of biodiversity
2. The sustainable use of biodiversity constituents
3. The equitable sharing of benefits from genetic resources [3, 8]

For these reasons, biodiversity is a general term covering species of all kinds of habitats and includes *genetic diversity*, *species diversity*, and *ecosystem diversity* [9–11]. Thus, loss of biodiversity covers much more than the future of individual species. *Genetic diversity* defines the difference between populations within a single species such as thousands of dog or rose breeds. Reducing genetic diversity within a species can result in the loss of its useful or desirable qualities (e.g., resistance to pests and diseases). Reducing genetic diversity can destroy the possibility of using these untapped resources as future food, technical or medicinal organisms [2, 9]. The term **species diversity** (between species) is associated mainly with the biodiversity, including various species of animals, plants, fungi, lichens, and microorganisms. In particular, endemic species, i.e., species limited to one geographic area, are important for global biodiversity given their unique genetic structure, and their extinction will mean an even greater loss of other species [9, 10]. Various human activities such as overexploitation of resources, pollution, and habitat changes affect the diversity of species. These factors lead to a gradual disappearance of species at all levels of habitats. In addition, the unreasoned introduction of species into new ecosystems often leads to a disturbance of the natural balance in those particular ecosystems [12]. Therefore, the constant growth of tourism, intensive agriculture (e.g., the practicing of monoculture), and the rapid development of industries and transport cause significant negative trends [4, 5]. Global climate change and population growth also have an adverse impact on the diversity of species. Loss of biological diversity can take many forms, but the most dramatic of these is the extinction of species. Each species has its own time on the geological clock, so the extinction of the species is a natural process that takes place without the participation of a person. However, it is impossible to challenge the fact that the extinction of species as a result of human activities far exceeds the rates of natural extinction [1, 13, 14]. *Ecosystem biodiversity*, a component of biodiversity, provides the basic necessities of life and shapes the human cultures. Besides those provisions, ecosystems also support and maintain life processes such as production of biomass and nutrient cycling (ancillary services), which are essential for human well-being [15, 16]. For example, coral reefs are not only rich in species diversity but also provide various benefits such as providing nutrients to people, protecting from storms, and recreation. However, according to various studies, it is estimated that 90% of coral reefs will be under threat of extinction by 2030 due to negative human activities [17].

4.2 Loss of Biodiversity

Global Biodiversity Outlook 3 (GBO-3) concluded that significant threats on biodiversity were increasing. These threats were defined as loss of biological diversity and mainly attributed to the unprecedented human impact on nature [18]. The main factors responsible for the loss of biodiversity include [8]:

- Destruction of natural habitats
- Introduction of invasive alien species
- Overuse of natural resources
- Pollution, in particular, the accumulation in the environment of nutrients, including nitrogen and phosphorus
- Greenhouse effect (climate change and acidification due to the accumulation of greenhouse gases in the atmosphere)

The abovementioned factors disturbed or even destroy natural habitats, causing a decrease in the genetic diversity of species and their extinction. The pace of economic development and the associated impacts on the living environment of living organisms are accelerating more and more [8].

4.2.1 Destruction of Natural Habitats

Mass destruction and fragmentation of habitats for agriculture, urban development, transport, tourism, forestry, and mining are the biggest threat to biodiversity [5, 9, 10, 19]. According to the International Union for Conservation of Nature (IUCN), habitat destruction affects more than 85% of the amphibians, mammals, and all birds, under threat of extinction [17]. For example, loud noise during transportation or presence of people who often behave very noisily disturbs wild animals and also leads to the fact that birds are afraid to fly up to their nests, not to unmask the chicks, and because of this, chicks are killed by hunger or hypothermia. Population growth and resource consumption is also an important factor in destruction of habitats. Moreover, the development of tourist activities often focuses on the maximum increase in the number of tourists visiting the area during one season. For this purpose, large hotel complexes are being built, destroying natural areas and creating sources of domestic garbage and sewage. All these social and industrial activities in nature lead to the fact that much more effort is required to restore populations of wild plants and animals, and the probability of loss of biodiversity is increasing [3–5]. For example, the natural habitat and the feeding areas of some animals are directly affected by deforestation, vegetation, and topography changes due to the construction of airports. In addition, it may cause behavioral changes due to physiological effects in various animals, especially farm animals within the airport vicinity [20, 21].

4.2.2   Introduction of Invasive Alien Species

Nonindigenous species, known as invasive species or exotic species in an ecosystem, damage the species present naturally inhabiting that system [22]. For this reason, biodiversity loss is accelerated by the intentional and unintentional introduction of invasive species. Increased global trade, transport and tourism, has expanded the introduction of invasive or exotic species to new environments. All taxonomic groups, such as viruses, fungi, algae, mosses, ferns, higher plants, invertebrates, fish, amphibians, reptiles, birds, and mammals, have alien invasive species [8, 23]. Species that have been intentionally or accidentally introduced into new environments have contributed to more than half of all animal extinctions that are known to occur. The types of species also entail huge economic costs. Annual economic damage from invasive alien species in crop production, fishing, and other areas is estimated at billions of dollars. The impact of these species on biodiversity in some cases is also extremely large and practically impossible to assess [24–26].

4.2.3   Overexploitation of Biological Resources

Overexploitation of biological resources leads to their depletion, especially in the field of renewable resources: forests, fish, and farm animals. Such use can in part be attributed to high population densities in some areas of the Earth, as well as to the ever-growing need of mankind for biological resources and the development of international trade [9, 27].

4.2.4   Pollution

Soil, water, and air pollution are the result of certain human activities, such as energy and industry, as well as excessive use of mineral fertilizers and pesticides in agriculture. Therefore, pollution is one of the factors causing the decrease in hundreds of species worldwide [8, 10]. For example, pollutants released into the atmosphere lead to the formation of acid rain, which adversely affects more fragile and porous newly laid bird eggs [28]. For example, air, water, soil, noise, nonionizing, visual, and aesthetic pollution are considered the environmental pollution factors due to airport activities [29]. In particular, aircraft landing and takeoff, engine braking after landing, ground maneuvers, and noise during engine testing [29] are the most important aviation activities that negatively affect biodiversity [21, 30, 31]. Aircraft noise does not only affect human health but may also cause the animal communities in the region to be frightened and migrate, thus changing their settlements [20, 21]. For this reason, environmental impact assessment (EIA) is an important tool for controlling environmental pollution effects resulting from the presence of airports in the region and ensuring sustainable development [20, 21, 30].

4.2.5   Global Warming

Global warming is also a serious threat to biodiversity around the world. As atmospheric CO_2 increases, global average temperature will continue to increase, and so, many habitats of plants and animals will change, depriving their inhabitants of habitual habitats and ecological niches to which these species are adapted. Furthermore, climate changes are not spatially homogeneous [32]. For example, the Arctic area experiences much larger changes due to global warming, while some locations are exposed to secondary effects like sea level increase [33]. Therefore, climate change and ocean acidifications may have already resulted in several recent species extinction. Some habitats, especially coral reefs, rivers, and mountains, are significantly affected by these climatic pressures [34, 35]. One of the important sources for greenhouse gases that cause climate changes is the aviation activities [21]. For this reason, the "Airport Carbon Accreditation" initiative was launched at the ACI EUROPE 2009 meeting to evaluate and recognize the efforts of participant airports to reduce and manage CO_2 emissions [36].

4.3   Biodiversity Conservation

GBO-3 concluded that biodiversity loss might decrease over time, if governments and society take coordinated action at different levels [8, 15, 36]. This meant understanding and minimizing the pressures on biodiversity and ecosystems. Hence, there are three main reasons for conserving biodiversity:

• Biodiversity brings both economic and scientific benefits for a person or can be useful in the future (e.g., in the search for new drugs or treatments).
• Biodiversity conservation is an ethical choice for the people because mankind is a part of the ecological system of the planet and has to take care of the biosphere.

- Conservation of biodiversity is an eternal, essential, and enduring value for humankind [2–5, 10].

For these reasons, it is very important to fulfill the three tasks given below in the sphere of biodiversity conservation [10]. These are:

1. *Economic* – biodiversity should be included in the country's macroeconomic indicators to provide direct (medicine, raw material and pharmacy, etc.) and indirect (ecotourism) economic benefits.
2. *Management* – the creation of partnership by involving in joint activities of state and commercial organizations, the army and navy, non-state organizations, the local population, and the general public should be required for the conservation of biodiversity.
3. *Legal* – it is necessary to include relevant terms and concepts in the legislation to establish legal support for the conservation of biodiversity.
4. *Scientific* – several scientific activities such as a formalizing decision-making procedures, search for biodiversity indicators, compilation of biodiversity cadasters, and organization of monitoring should be planned for the conservation of biodiversity.

According to Global Biodiversity Strategy, conservation of biodiversity in the long term can be a sustainable process only if the concern of the society and its belief in the need for action in this direction will constantly increase. Consequently, all living organisms are important for the people, and so biodiversity management is necessary for the sustainable use of biodiversity [8, 37].

4.4   Biodiversity Management and Methods

Strategic Plan for Biodiversity 2011–2020 agreed at the tenth meeting of the convention of biological diversity (CBD) conference of the parties (COP 10) in Nagoya, Aichi Prefecture, Japan, in 2010. Until 2020, Aichi Biodiversity Targets (20 targets) have been accepted as a strategic plan for the biodiversity and sustainable use targets all over the world [4, 8]. The Strategic Plan includes five interdependent strategic goals (A–E) and 20 targets (T1–T20). These are classified in the Global Biodiversity Outlook 4 (GBO-4) as follows [4, 8]:

Goal A: Main reasons or indirect factors of loss of biodiversity, including lack of knowledge about biodiversity and its values. This goal includes four sub-goals, indicated as following targets:

- **T1**: Biodiversity awareness will be increased by 2020, at the latest.
- **T2**: Biodiversity values will be integrated into all processes by 2020, at the latest.
- **T3**: Incentives will be reformed by 2020, at the latest.
- **T4**: Sustainable production and consumption will be planned by 2020, at the latest.

Goal B: Reduce the direct pressures on biodiversity and promote sustainable use. This goal includes six sub-goals, indicated as following targets:

- **T5**: The loss of all habitats will be halved or will be reduced by 2020, at the latest.
- **T6**: Sustainable management of aquatic living resources will be fully implemented by 2020, at the latest.
- **T7**: Sustainable agriculture, aquaculture, and forestry will be implemented by 2020, at the latest.
- **T8**: Pollution will be reduced by 2020, at the latest.
- **T9**: Invasive alien species will be prevented and will be controlled by 2020, at the latest.
- **T10**: Ecosystems vulnerable to climate change by 2015.

Goal C: Improve the status of biodiversity by safeguarding ecosystems, species, and genetic diversity. This goal includes three sub-goals, indicated as following targets:

- **T11**: At least 17 percent of terrestrial and inland water areas and at least 10 percent of coastal and marine areas will be conserved by 2020.
- **T12**: Reducing risk of extinction by 2020.
- **T13**: Safeguarding genetic diversity by 2020.

Goal D: *Improve* benefits for all biodiversity and ecosystem services. This goal includes three sub-goals, indicated as following targets:

- **T14**: Ecosystem services by 2020
- **T15**: Ecosystem restoration and resilience by 2020
- **T16**: Accessing and sharing benefits from genetic resources by 2015

Goal E: *Improve* implementation through joint planning, knowledge management, and capacity building. This goal includes four sub-goals, indicated as following targets:

- **T17**: Biodiversity strategies and action plan.
- **T18**: Traditional knowledge, practices, and innovations will be integrated by 2020.
- **T19**: Sharing information and knowledge by 2020.
- **T20**: Mobilizing resources from all sources by 2020.

As mentioned in Aichi Biodiversity Targets, sustainable management of biodiversity is one of the most important issues on the world agenda. Biodiversity management is not an easy task. In particular, the fact that the relevant groups cannot be easily affected is valid in this process. Most of the time, limited amounts of resources are allocated for the international agreements and regulations of biodiversity. Relevant groups include government agencies, business structures, nongovernmental organizations, and the local population. Some changes, such as climatic changes, have unpredictable consequences for the conservation of biodiversity and thus represent factors that complicate the management of biodiversity [4]. Despite all the difficulties, biodiversity management includes special stages to formalize and structure the management process. These stages are definition of the problem (1), planning (2), monitoring of biodiversity (3), and evaluation (4). According to these stages, biodiversity management has the following objectives [10]:

- Plan and coordinate all efforts to conservation of biodiversity.
- Conserve and restore ecosystem, species, and genetic diversity using in situ and ex situ conservation strategies.
- Use biodiversity in sustainable management systems such as forestry, fisheries, livestock, and agriculture.
- Share the benefits of biodiversity equally with social and cultural means.
- Provide a legal framework for preservation and sustainable use of biodiversity.
- To create opportunities for people and organizations to integrate environmental measures in a bioregional scale into plans and schemes for social and economic development.

4.4.1   Stage 1: Definition of the Problem

On a global scale, definition of the problem of biodiversity managements is associated with ever-increasing degradation in nature (the number of populations, species, communities, ecosystems, and their state, i.e., wildlife in general). Often this problem is understood or denoted at the policy level more narrowly – as the problem of the direct human impact on wildlife. It is very important to understand this problem as one of the important problems facing humanity [1]. For example, it is stated that, when airport activities are not carried out in accordance with environmental management programs, they may have negative effects on the plants and animals living in the area. It is suggested that the ecological effects of airports emerged as a result of construction work and daily activities in the airports. For example, the use of reserved green habitats for airport construction can lead to vegetation changes, the migration of animals living in the region, and the deterioration of ecological balance [20, 21, 29, 30]. For this reason, determining environmental problems related to ecosystem and species diversity is very important in the sustainable management of biodiversity.

4.4.2   Stage 2: Planning of Biodiversity Management

The expected results of the management policy are closely related to the management concepts used. In particular, there are two general plans in biodiversity management, namely, ecosystem and species management, and each of them gives different results. Management plans including species management, classification and conservation of species, and integration with economic aspects are part of biodiversity management [10].

Species conservation plans can be developed at the local and global levels. A number of countries already have relevant laws, which create the legal basis for sustainable management of conservation of species. Plans for the restoration of species are special plans, usually developed within the framework of legislation, with the assignment to different species of a particular status. Typically, such a plan aims to restore the species to such an extent that it has moved from the category of endangered species to the category of species not subject to the threat of extinction. The Species Survival Commission (SSC) has established threat criteria for the classification of species nearing extinction [24, 38]. Taxa that meet the criteria for inclusion in one of the threatened categories are then included in the Red Lists and Red Books [17]. The purpose of such actions is to create a database of species that are threatened to extinction, develop a basis for identifying priority conservation and protection objects, and monitor the effectiveness of measures to restore the number of species. According to the Red List of Threatened Species, 8417 species are threatened including plants (2443 species), invertebrates (2930 species), marine fishes (351 species), freshwater fishes (115 species), mammals (649 species), reptiles (1175 species), and birds (238 species). These assessments indicate a decrease in many species worldwide [9, 38–41]. Therefore, research on the planning of species distribution in habitats is very important for the development of sustainable biodiversity management in special areas such as airports. Nowadays, planning, construction, and operation of airports considering environmental sustainability and biodiversity management is a necessary approach [21, 29, 30]. In particular, significant efforts are being made to maintain biodiversity at airports, with the preservation of flora and fauna. For example, Gatwick Airport in London has developed 5-year biodiversity management action plans for 2012–2017. These biodiversity actions focused on three on-airport sites and increased the biodiversity value of the airport and the educational value of biodiversity. Moreover, over the Biodiversity Action Plan period, a comprehensive database of all species recorded so far at Gatwick Airport has been created and several of which are notable and of conservation purposes [42]. An important issue in the determination of sustainable biodiversity management programs in aviation is the bird activities around the airport and migration routes along the flight route. In addition, flight activity of birds is a potential danger in terms of flight safety. For this reason, the general principle is to prevent birds from entering the boundaries of the airport. Airport authorities should develop and implement wildlife and bird control programs, taking into account the regional characteristics of the airport. The programs should cover:

- The methods to be applied
- The personnel to be employed
- Personnel training
- The tools, equipment, and materials to be used
- The organizations providing services at the airport
- The coordination and cooperation with the local authorities and institutions [43]

The International Bird Strike Committee (IBSC) published the document titled "Standards for Aerodrome Bird/Wildlife Control, 2006," which includes standards for the control and removal of birds and other living organisms at risk in airports [21, 44]. The document emphasizes that the standards required for the efficient functioning of programs for bird and wildlife control should be met at all airports [21, 44]. There are several handbooks at the airports that explain the techniques that can be used to manage the risk of collision with the birds during landing and departure of the aircraft and the risk of striking wildlife at the airport. Several studies stated that methods of minimizing bird food sources and reducing attractiveness of some tree species at airports for birds have become widespread and these studies have been carried out in cooperation with local administrations [21, 44–47].

4.4.3   Stage 3: Monitoring of Biodiversity

Monitoring is an effective tool for the biodiversity management. This method is used to measure the effectiveness of activities undertaken to conserve biodiversity and to identify biological trends, both natural and anthropogenic. Therefore, environmental monitoring studies provide for evaluation of abiotic (i.e., chemical parameters of air, water, soils, etc.) and biotic changes (i.e., changes in composition, abundance, distribution of species, and communities) in the ecosystem [10]. "Environmental monitoring" means the measurement of environmental parameters (indicators) at regular intervals over a long period of time. Environmental monitoring evaluates environmental and biological changes within an ecosystem in order to separate natural fluctuations from abnormal changes. For example, abnormal variations in the number of species can be observed due to catastrophic causes (i.e., the death of animals and plants under the influence of pollution, oil spills, etc.) [10, 48]. Monitoring also helps discover the cause-effect relationships between external factors and changes in species populations or in ecosystems as a whole. Environmental monitoring is important in the inventory of mutual relations of species and habitats in the region where the assessment is conducted [49, 50].

In a monitoring program over a certain period of time, several methodical approaches such as "All biota Taxonomic Inventory (ABTI)," "All Taxa Biodiversity Inventory (ATBI)," and "A Rapid Biodiversity Assessment (RBA)" can be applied [10, 50]. ABTI focuses on several key groups of species that scientists believe are good indicators of the status of particular ecosystems (e.g., termites, fish, or butterflies). ATBI aims at describing all species present in a certain area. RBA bases on lists of selected species that provide an operational indicator of the biological richness of an area. Further, monitoring stage of biodiversity management focuses on the diversity of genetic, species, or habitats or combinations thereof [10, 50]. However, the approach used in the work can take time to identify clear patterns and trends in all direction due to the natural spatial and temporal variability of ecological systems and the complexity of the relationships between ecosystem components as well as factors such as the weather and seasonal conditions [6]. Monitoring of biological diversity at the local, regional, national, and global levels requires a systemic and environmental infrastructure, economic support, and human resources. For this reason, the selected method is closely linked to the aim of the monitoring program and the resources that can be made available, which can be thought of as cost effectiveness [10, 50, 51].

At this stage of biodiversity management, environmental and social impact assessment, an accepted technical research method, can be used. Because environmental and social impact assessment (ESIA) is a major approach to identifying serious problems before biodiversity threats manifest themselves [50, 51]. Therefore, the ESIA is a useful tool for predicting and evaluating the potential environmental and social impacts of projects evaluating alternatives and developing appropriate mitigation, management, and monitoring measures. The most important stage of such an assessment is a survey of the selected area. Environmental impact assessment is a strategically important legal tool for the protection of biodiversity, as it aims to eliminate problems before the implementation of projects. For most countries, the ESIA is a legal requirement for all new developments and important changes to the work on the site [50, 51]. Such an assessment should be carried out within the framework of individual industries, land use types, programs, and plans: in particular, when planning the construction of highways, changes in the river basin water regime, forest management, etc. [21, 50]. If the project has already become an integral part of the approved plan or program, it is often too late or impossible to do such an assessment at the implementation stage to prevent major damage. In light of this information, when airports are evaluated, it is known that direct landing takeoff and air navigation of aircraft direct and indirectly affect biodiversity [21]. For example, according to the major bird migration road map worldwide, Turkey is an important migration country [52]. When the migration route map of Turkey was examined, one of the main migration routes that coincide with the Istanbul-Antalya-Adana-Hatay route seemed to have the busiest air traffic. Approximately, 60% of Turkey's air traffic is concentrated in areas where this route passes, and this poses potential risks with respect to flight safety and environmental dimensions. For this reason, periodic bird watching is carried out every year at Hatay airport as a requirement of the EIA report [31, 43].

Therefore, sustainable environmental management programs in airports today are becoming increasingly widespread [21]. These programs can minimize the abovementioned negative effects of airport activities by implementing integrated environmental management techniques. Generally environmental sustainability practices in airports include monitoring and/or measuring water conservation, water quality, climate change, air quality, land use, biodiversity, environmentally sustainable materials, waste, noise, aesthetics, energy, and green buildings [53]. For these reasons, biodiversity is monitored in sustainable management programs at airports [21]. In these monitoring studies, preservation of ecological balance is the main emphasis, and thus EIA reports in rural areas around the airports should include the following efforts [21, 31]:

• Management of green areas as meadow areas
• Monitoring of fauna and flora structure at given times

- Determination of endemic species
- Annual monitoring of population density of indicator organisms
- Creation of new habitats by afforestation in remote areas for bird species
- Determining and monitoring migration routes of animals
- Preventing attractive environments for birds that present danger to aircraft

4.4.4   Stage 4: Evaluation

At this stage, it is necessary to check whether biodiversity management is sustainable or not. For this reason, it is necessary to answer the problems determined in the management stages, to solve the problems in the study area, and to determine what needs to be improved in the future [10, 48]. Therefore, at the end of the monitoring studies, biodiversity status for new habitats in a given area should be evaluated and made sure that adequate measures are in place to secure the sustainability of this. Finally, annual reports including all management processes and results should be prepared and published [48, 51]. For example, a review was published by Rachel Bicker, biodiversity consultant at Gatwick Airport, at March 2018, and it detailed the outcomes over 5 years (2012–2017) of biodiversity management at Gatwick Airport toward the Biodiversity Action Plan (BAP) objectives and targets [42]. This report includes monitoring of outcomes, all species summaries, biodiversity area maps, biodiversity awards, studies of the volunteering and community engagement, habitat condition analysis, and biodiversity performance indicator assessment criteria. Therefore, it was reported that biodiversity management process in Gatwick Airport provided an overview of the work undertaken, the project achievements, and areas of future focus. In light of this previous review, the results from reports may contribute to the development of new biodiversity management action plans [42]. For this reason, the new management plans should address any issues that have been identified according to the results of previous reports and continue to provide the sustainability for biodiversity management at the airports.

4.5   General Approaches for Biodiversity Management at Airports

The activities implemented for the purpose of biodiversity and natural life conservation during environmental sustainability practices at international airports vary in different regions (Europe, Asia, Canada, the UK, and the USA (wide network)). These activities are listed below [21, 31, 43, 53]:

- Equipment to frighten birds with sound systems in airports.
- Arrangements according to the need for water tanks on water treatment systems.

- Greening in more than half of the airport, operationally 80% and 20% in nature conservation zones, forest, and water bodies – annual plan applied by green area maintenance service, annual monitoring according to "The Provisions of The Animal Nutrition Administration."
- Preventing bird strikes (with airplanes), managing extensive lawn areas in airport open spaces – supporting other nature conservation areas and promoting the development of important ecological zones that act as habitat for endangered flora and fauna.
- Financial ecological/compensation measures supported and created by the airport.
- Dolphin shelter/marine park, The Committee for the Protection of the Marine Mammals of Endangered Species.
- Use of crackers to scare birds.
- Support the local green safety belt and outreach program, which includes improving staff and habitat.
- Protect and improve coastal habitat during bent repair.
- To finance nature protection to manage the local nature reserve.
- Nonlethal techniques for bird control, life management, monitoring, birds' removal of aircraft (such as noise makers, dogs).
- Regional cooperation.
- Wildlife Hazard Management Plan.
- Construction management.
- Wetlands breeding program.
- Research support.
- Vegetation management/habitat protection. Sand dune protected area – the largest coastal sand dune in Southern California, hosted by the largest coastal dune piece, monitoring the health of endangered species on an annual basis with land surveying technicians and biologists, removing environmental pests.
- Restoration of grassland habitats with common fund with local conservation group.
- Contract with the US Department of Agriculture.
- The wildlife management plan pay for land clearance permission (Clean Water Act).
- Transport of threatened species to the conservation areas.
- A new landscaping program designed to minimize birds.

References

1. Sala OE, Chapin FS, Armesto JJ, Berlow E, Bloomfield J, Dirzo R, Huber-Sanwald E, Huenneke LF, Jackson RB, Kinzig A, Leemans R, Lodge DM, Mooney HA, Oesterheld M, Poff NL, Sykes MT, Walker BH, Walker M, Wall DH (2000) Biodiversity – global biodiversity scenarios for the year 2100. Science 287:1770–1774
2. Sala OE (1995) In: Mooney HA, Lubchenco J, Dirzo R, Sala OE (eds) Global biodiversity assessment, section 5. Cambridge University Press, Cambridge

3. Biodiversity – GreenFacts. Available at: https://www.greenfacts.org/en/digests/biodiversity. htm. Accessed 10 Oct 2018
4. Secretariat of the Convention on Biological Diversity (2014) Global Biodiversity Outlook 4. Montréal, 155 pages. Available at: http://www.cbd.int/sp/targets/. Accessed 10 Oct 2018
5. Loss of biodiversity and extinctions—global issues. Available at: http://www.globalissues.org/article/171/loss-of-biodiversity-and-extinctions. Accessed 10 Oct 2018
6. Finlayson CM, Everard M, Irvine K, McInnes R, Middleton B, van Dam A, Davidson NC (eds) (2018) The wetland book I: structure and function, management, and methods. Springer Netherlands, Dordrecht
7. The Biodiversity Indicators Partnership (BIP) (2018). Available at: https://www.bipindicators. net. Accessed 10 Oct 2018
8. The Convention on Biological Diversity (CBD). Available at: https://www.cbd.int/cop10. Accessed 10 Oct 2018
9. Conservation International. Biodiversity hotspots. Available at: http://www.conservation.org/where/priority_areas/hotspots/Pages/hotspots_main.aspx. Accessed 10 Oct 2018
10. Bat L, Sezgin M, Toplaoğlu B, Ivanova M, Antonova B (2018) Biodiversity module. Available at: http://www.coastlearn.org/biodiversity. Accessed 10 Oct 2018
11. Group on Earth Observation Biodiversity Observation Network (2010) GEO BON detailed ımplementation plan version 1.0. Available, Geneva, Switzerland. Available at: http://www.earthobservations.org/documents/cop/bi_geobon/geobon_detailed_imp_plan.pdf.   Accessed 10 Oct 2018
12. Primack RB (1998) Essentials of conservation biology. Sinauer Associates, Sunderland MA, p 260
13. Thomas CD, Cameron A, Green RE, Bakkenes M, Beaumont LJ, Collingham YC, Erasmus BFN, de Siqueira MF, Grainger A, Hannah L, Hughes L, Huntley B, van Jaarsveld AS, Midgley GF, Miles L, Ortega-Huerta MA, Peterson AT, Phillips OL, Williams SE (2004) Extinction risk from climate change. Nature 427:145–148
14. IPCC (2007) Climate change 2007 – impacts, adaptation and vulnerability. Cambridge University Press, Cambridge
15. Castro P et al (eds) (2016) Biodiversity and education for sustainable development, World sustainability series. Springer International Publishing, Cham. https://doi.org/10.1007/978-3-319-32318-3_1
16. Haines-Young R, Potschin M (2010) The links between biodiversity, ecosystem services and human well-being. In: Raffaelli DG, Frid CLJ (eds) Ecosystem ecology. A new synthesis. Cambridge University Press, Cambridge, pp 110–139
17. The IUCN red list of threatened species. Available at: www.iucnredlist.org. Accessed 10 Oct 2018
18. Global Biodiversity Outlook-3, 2010. Available at: https://www.cbd.int/gbo3/. Accessed 10 Oct 2018
19. Vié J-C, Hilton-Taylor C, Stuart SN (eds) (2009) Wildlife in a changing world – an analysis of the 2008 IUCN red list of threatened species. IUCN. 180 pp, Gland
20. Korul V (2004) Havaalanı Çevre Yönetim Sistemi. SHYO Sosyal Bilimler Dergisi:99–120. (in Turkish)
21. The Aviation Environment Federation (AEF) What are an airport's impacts? Available at: https://www.aef.org.uk/issues/biodiversity/. Accessed 10 Oct 2018
22. Drake JA et al (1989) Biological invasions: a global perspective. Wiley, Chichester, p 37
23. Roy HE, Roy DB, Roques A (2011) Inventory of terrestrial alien arthropod predators and parasites established in Europe. BioControl 56:477–504
24. MITSG (1996) Massachusetts Institute of Technology Sea Grant College Program. Abstracts from the exotic species workshop: issues relating to aquaculture and biodiversity, MITSG, Cambridge, MA, 57 pp
25. IUCN guıdelines for the preventıon of biodiversity loss caused by alien invasive species. Prepared by the SSC Invasive Species Specialist Group, Approved by the 51st Meeting of the IUCN Council, Gland, February 2000
26. Invasive Species Specialist Group (ISSG). Available at: http://www.issg.org/index.html. Accessed 1 Oct 2018

27. The International Union for Conservation of Nature (IUCN). Available at: https://www.iucn.org/our-work. Accessed 1 Oct 2018
28. Naturopa C (1996) Biodiversity: questions and answers. Council of Europe Publishing, Strasbourg., F-67075 Strasbourg Cedex
29. Whitelegg J, Williams N (2000) The plane truth: aviation and the environment. Transport 2000 and The Ashden Trust, London
30. ACI (2010) Airports and environments ACI position brief. Washington, DC, ACI
31. Oto N (2011) Çevresel sürdürülebilirlik havaalanları: Esenboğa Havalimanı Örneği, Doktora Tezi, Ankara Üniversitesi, Sosyal Bilimler Üniversitesi, Ankara (*in Turkish*)
32. Nicole E, Heller E, Zavaleta S (2009) Biodiversity management in the face of climate change: a review of 22 years of recommendations. Biol Conserv 142:14–32
33. IPCC (2007) Climate change 2007 – the physical science basis. Cambridge University Press, Cambridge
34. McLaughlin JF, Hellmann JJ, Boggs CL, Ehrlich PR (2002) Climate change hastens population extinctions. Proc Natl Acad Sci U S A 99:6070–6074
35. Pounds JA, Bustamante MR, Coloma LA, Consuegra JA, Fogden MPL, Foster PN, La Marca E, Masters KL, Merino- Viteri A, Puschendorf R, Ron SR, Sanchez-Azofeifa GA, Still CJ, Young BE (2006) Widespread amphibian extinctions from epidemic disease driven by global warming. Nature 439:161–167
36. ACI EUROPE (2010) Airport carbon accreditation: annual report 2009–2010. ACI EUROPE, Brussel
37. Alves F, Filho WL, Araújo MJ, Azeiteiro UM (2013) Crossing borders and linking plural knowledge: biodiversity conservation, ecosystem services and human well–being. Int J Innov Sustain Dev 7(2):111–125
38. International Union for Conservation of Nature (IUCN) (2017) Annual report, 2017
39. Del Mar Otero M et al (2017) Overview of the conservation status of Mediterranean Anthozoa. IUCN, Malaga
40. Garcia Criado M et al (2017) European red list of lycopods and ferns. IUCN, Brussels
41. Pippard H et al (2017) The conservation status of marine biodiversity of the Pacific Islands of Oceania. IUCN, Gland
42. Biodiversity Consultant (2018) Gatwick biodiversity action plan five year review. 2012–2017 report by Rachel bicker. Gatwick Airport, Mar 2018
43. General Directorate of State Airports Authority (DHMI). (2017) Annual report, 2017. Available at: https://www.dhmi.gov.tr
44. The International Bird Strike Committee (2006) Recommended practices no. 1, standards for aerodrome bird/wildlife control, issue 1, Oct 2006
45. ACI (2005) Aerodrome bird hazard prevention and wildlife management handbook, 1st edn. Airports Council International, Geneva
46. CAA (1998) CAP 680 bird control on aerodromes. Civil Aviation Authority, London
47. Cleary EC, Dolbeer RA (1999) Wildlife hazard management at airports, a manual for airport personnel. US Federal Aviation Administration, Washington, DC
48. CSI (Cement Sustainability Initiative) (2014) Biodiversity management plan (BMP) guidance. Available at: https://docs.wbcsd.org/2014/09/CSI_BMP_Guidance.pdf. Accessed 18 Feb 2019
49. Convention on Biological Diversity. National biodiversity strategies and action plans. Available at: http://www.cbd.int/nbsap/. Accessed 1 Oct 2018
50. Treweek J (2009) Ecological impact assessment. John Wiley & Sons, Chichester. isbn:1444313290, 978144431329
51. CSI (Cement Sustainability Initiative) (2005) CSI, environmental and social impact assessment guidelines. Available at: http://www.wbcsdcement.org/pdf/cement_esia_guidelines.pdf. Accessed 1 Oct 2018
52. Turan L, Dengiz Ş, Yüksel E, Ertaş C (2009) Ankara Esenboğa Havalimanı Vahşi Hayat ve Kuşla Mücadele Çalışması. İnforama, Ankara. (*in Turkish*)
53. ACRP (2008) ACRP synthesis 10 airport sustainability practices, a synthesis of airport practice. TRB, Washington, DC

Chapter 5
A Holistic View of Sustainable Aviation

Birol Kılkış

This chapter redefines the concept of sustainable aviation (SA) in a holistic model that encompasses the landside and the airside of aviation. It also covers the type of the aviation fuel, ground transport, safe utilization of renewable energy resources near airports, waste management, biogas, new HVAC technologies, aviation and passenger safety, energy, and exergy economy, all being mapped on a new circular exergy diagram. This chapter gives special emphasis to terminal buildings and presents a new exergy-based rating system. Three international airports were compared in terms of CO_2 emissions, sustainability, and rational utilization of energy and exergy.

5.1 Introduction

Sustainable aviation needs to cover three major factors simultaneously without any compromise among them, and thereby, all definitions, rating metrics, sustainable design, and operation along with the economy must be collectively based on them. These three factors are:

Airside: Aircraft flight including taxiing, takeoff, descent, landing and parking, aviation fuel, air traffic optimization

Landside: Terminal buildings, ground handling, maintenance, repair and overhaul (MRO), terminal ground transport, commuting transport, heating, ventilating and air-conditioning (HVAC) systems, and energy infrastructure

Environment: Environmental impact, carbon footprint, and economy

B. Kılkış (✉)
Turkish HVAC&R Society, Bestekar Cad. Kavaklidere 06680 Ankara, Turkey

© Springer Nature Switzerland AG 2019
T. H. Karakoc et al. (eds.), *Sustainable Aviation*,
https://doi.org/10.1007/978-3-030-14195-0_5

5.2 Airside

Airside operations for sustainability mainly depend upon the fuel economy and associated global warming effect. According to the United Nations Intergovernmental Panel on Climate Change (IPCC), aviation produces around 2% of the world's manmade emissions of carbon dioxide (CO_2).

As aviation grows to meet the increasing demand, particularly in fast-growing emerging markets, the IPCC forecasts that its share of global manmade CO_2 emissions will increase to around 3% in 2050 [1, 2]. Although the aviation industry will be responsible for CO_2 emissions by an amount less than 5%, the aviation industry suffers from a large public distrust. Both for operational and environmental points of view, the industry must at any rate focus on clean fuel or "electric" planes, both of which seem to be far from complete reality with today's technology and economy. The electrical energy that may be used for charging the electric planes is subject to scrutiny by the origin of the primary fuel that is used at the first place, upstream of the chain. Battery risks and their weight are other major points of concern. Figure 5.1 shows that the exergy rationality, ψ_R, of a latest-technology jet engine is 0.38, which is a simple ratio of the supply and demand unit exergies. This value is slightly above the average of the built environment, which is close to 0.3.

On the other hand, the exergy rationality of an electric-driven aircraft propeller is 0.41 (Fig. 5.2). This minuscule increase means that the expected CO_2 emission savings are only about 10%. This comparison shows that unless the electric power is supplied by renewable energy systems and issues about electrical energy storage on board are solved, electric planes will not be truly feasible, except for light and

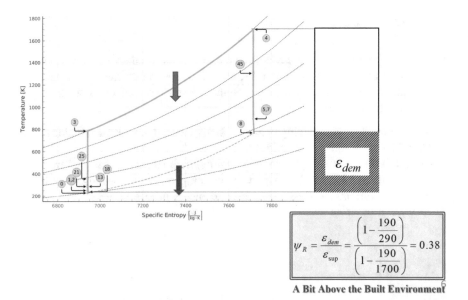

$$\psi_R = \frac{\varepsilon_{dem}}{\varepsilon_{sup}} = \frac{\left(1 - \dfrac{190}{290}\right)}{\left(1 - \dfrac{190}{1700}\right)} = 0.38$$

A Bit Above the Built Environment

Fig. 5.1 Exergy rationality for today's jet engine

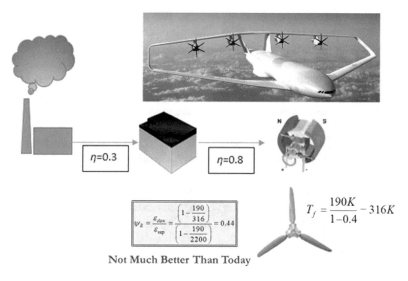

Fig. 5.2 How exergy rational is an electric plane with current technology

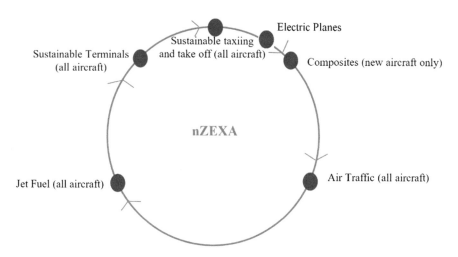

Fig. 5.3 Six steps in closing the gap for nZEXA and nZCA (nearly zero exergy aviation and nearly zero carbon aviation) [3]

small aircraft. Figure 5.3 shows the potential impact of electric planes in comparison to other factors on the circular exergy diagram [3]. This diagram shows that nearly- zero exergy aviation (nZEXA) that is destined to minimum exergy destructions, which leads to nearly- zero carbon aviation (nZCA), has six major components of decarbonization. Among them, the existing electric plane technology seems to have a potentially small impact.

Figure 5.3 shows that carbon-neutral jet fuel has a higher potential in the decarbonization process for aviation. Therefore, until the electric plane concept reaches new exergy-technology heights, carbon-neutral jet fuel may be the transitional solution until the electric planes of the future. According to a recent study, carbon-neutral or even carbon-negative jet fuel may be produced in an offshore combination of wind, wave, and solar energy (SWWCP) [4]. The electrical energy required for all chemical processes for producing jet fuel will be completely generated from such a system, for example, in the Black Sea where hydrogen sulfide (H_2S) gas is simultaneously extracted from the seabed [5]. Although the final jet fuel burnt in the jet engine will emit CO_2, it will be paid off by the harmful gases sequestered by the industry, which is explained below. So, the system will be carbon neutral. The Fischer-Tropsch (FT) process is a catalytic chemical reaction in which carbon monoxide (CO) and hydrogen (H_2) are converted into hydrocarbons of various molecular weights according to the following equation:

$$(2n+1)H_2 + nCO = C_nH_{(2n+2)} + nH_2O \quad \{n \text{ is an integer}\} \tag{5.1}$$

The hydrogen gas input to this process has two sources, namely, by direct electrolysis of (sea) water using electricity obtained from renewables at the SWWCP system and from the drilled seabed system of the same SWWCP system tapping the H_2S gas deposits. The following chemical process is used [6]:

$$2COS + O_2 = 2CO + SO_2 \tag{5.2}$$

CO may be fed back to reaction (5.1) if n is not too large or to other reactions. Carbonyl sulfide (COS) is found in many industrial process streams, such as those associated with petroleum refineries and coal gasification plants. Carbon disulfide (CS_2) is a greenhouse gas, which is associated with several industrial processes. This chemical is used to form CO and dry SO_2 [7]. Thus, an added value is generated, while greenhouse gas emissions are reduced in the industrial sector.

$$2CS_2 + 5O_2 = 2CO + 4SO_2 \tag{5.3}$$

$$CO + H_2S = H_2 + COS \tag{5.4}$$

Finally, H_2 is obtained from H_2S and COS is regenerated, which may be fed back to the reaction in Equation 5.3. Figure 5.4 shows the process model in brief where numbers in parenthesis are equation numbers.

5.3   Landside

Airports are becoming a vitally important part and commonplace of the everyday life and are getting more integrated with the neighboring cities. The presence of thousands of passengers at an internationally diverse scale is rapidly expanding,

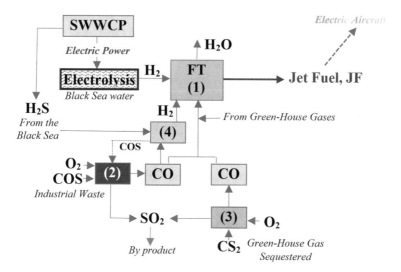

Fig. 5.4 Jet fuel from hydrogen from H_2S and industrial pollutants using renewable energy resources

which acts as potential breeding grounds for international epidemics at any instant without much notice. In this respect, hygiene is becoming a bigger concern besides human and property safety, especially in new very large airports. On the other hand, airports need to become much conscious about energy, environment, and service quality while operational costs are kept at a minimum level. In fact, several sustainability parameters do conflict with each other. The following list is a short punch card for a complete sustainability of an airport, which starts from the construction phase and continues through the operation of the airfield:

- Sustainable site development
- Water savings
- Energy efficiency
- Exergy rationality
- Materials selection
- Indoor environmental quality
- Hygiene
- Safety
- Waste management
- Utilization of renewable energy resources, including waste heat and biogas generation from airport wastes
- Optimally integrated operations
- Convenience and comfort both on the airside and the landside
- District energy (DE) system
- Low-exergy (LowEx) building

In 2014, San Diego Airport had just achieved LEED Platinum, the highest green building certification attainable. The airport, also known as Lindbergh Field, has it all: a megawatt of solar power, water and energy conservation measures, sophisticated climate and energy controls, a reflective roof, and improved indoor environmental quality through the use of low volatile organic compound adhesives, sealants, paints, and coatings. However, all these green building certification programs are focused more or less on the individual terminal buildings. They often ignore the external connections and relations of the airport with the built environment on the landside, let alone their vital connections on the airside. There are few exceptions like the Amsterdam Schiphol Airport that used to transfer its waste to the city for energy production and in turn receive power. With the advent of 4DH (fourth-generation district heating) systems in neighboring cities, airports have to get connected with them in order to put the term green one or more steps further. In this respect, the exchange of energy goes beyond electrical energy and begins to encompass heat and cold at different temperatures and energy forms. Figure 5.5 shows such a multi-energy transfer, which clearly marks that the quality of energy exchange (exergy) becomes more important than the quantity of energy exchange. In Fig. 5.5, an exergy exchange deficit occurs because the terminal delivers 30 °C water to the district but receives 40 °C water in the same amount over a given period of time. A similar deficit occurs for cooling because exchange temperatures of the chilled water are different. This makes it necessary to refer to the second law of thermodynamics [8–10]. Another approach is provided by a new airport ranking index developed in [11].

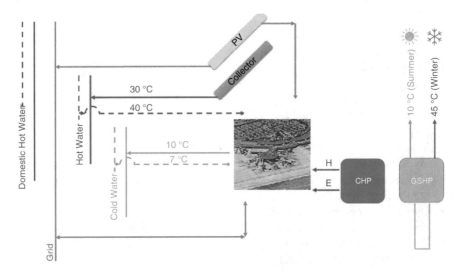

Fig. 5.5 Supply and return temperatures between a sustainable airport and DE system

5.4   Sample Airports

5.4.1   Copenhagen (CPH)

Copenhagen Airport (CPH) aims for its energy consumption to remain unchanged despite the projected growth. In order to meet this target, CPH is introducing energy-saving measures in order to accommodate the expected future growth in passenger numbers. First of all, the airport utilizes the energy obtained from the DE system of the city of Copenhagen. Furthermore, in 2014, CPH has focused on changing much of the airport's old ventilation equipment in Terminal 2 and in Pier C. The old equipment without heat recovery was replaced with a new ventilation system with "highly efficient" heat recovery units, which especially contributed to the large savings on heating. However, it must be noted here within the exergy context of this chapter that the system must be designed and operated according to the second law of thermodynamics. The following rule must govern the design such that the exergy of the thermal energy must be greater than the exergy demand of the ventilation system. If the outdoor temperature is T_o and the thermal preheating power provided by the exhaust air Q_H raises the temperature of the outdoor air to T_R, then the total ventilation power demand P_F is strongly restricted:

$$P_F \cdot (0.95) < Q_H \cdot \left(1 - \frac{T_o}{T_R} \right) \tag{5.5-a}$$

Here (0.95 W/W) is the unit exergy of electrical power. For example, if T_o is 283 K and T_R is 297 K, then P_F must be less than 5% of the thermal power recovered for indoor ventilation in an airport. This constraint is especially important in large terminal buildings, yet it is often ignored in design and application. If P_F exceeds this condition, then compound CO_2 emissions will be more than anticipated by the first law of thermodynamics. This excess emission depends on the primary source of electricity. According to CPH, their savings accounted for more than one-third of the overall energy savings. Yet, according to Equations 5.5-a, and 5.5-b, it is most likely that exergy savings will be none existant or negative:

$$COPEX = COP \cdot \frac{\left(1 - \dfrac{T_o}{T_R} \right)}{0.95} = \frac{Q}{P_F} \cdot \frac{\left(1 - \dfrac{T_o}{T_R} \right)}{0.95} \tag{5.5-b}$$

The exergy-based *COP*, namely, *COPEX*, will only be a small fraction of *COP*, which takes place in the heat recovery system catalogs and manuals in the sector.

A similar condition applies to heat pumps, which are widely used in airports today. Heat pumps are driven by electric power and generate heat with *COP* values around 4 or 5, depending upon the supply and demand temperatures, the thermal source (ground, air, or water), and the type of heat pump. If *COP* is 4 and the above numerical example applies with $T_o = 310$ K and T_R is 330 K on the supply side of the heat pump, then *COPEX* will only be 0.25. In order to have a break-even condition, *COP* needs to be 15.7, which is practically impossible unless the temperature of the source and the demand temperature of the HVAC system are brought very close to each other. This is possible through the implementation of LowEx buildings.

The new system at CPH, however, contributed to an improved indoor climate and provides the ability to control the system as needed. Groundwater cooling system (ATES) has led to a 30% increase in cooling capacity. Hereby the entire Pier B and Pier C and parts of Terminal 2 are being cooled by this system. In 2014, CPH's management adopted an updated environmental and climate policy. This updated policy includes guidelines for CPH's work with the environment, climate, and energy, all detailed in a single policy. The environmental principles from the UN Global Compact and basic principles for the work with environmental management are now integrated more clearly in the policy.

5.4.2   Schiphol, Amsterdam

Schiphol Airport has several green features, which were especially implemented after a big expansion and upgrading effort. It incorporates solar energy, cogeneration, absorption cooling, energy storage, district connection, and a comprehensive waste management and energy recovery. Assume that this airport competes with other airports for becoming the major hub for Europe with the primary advantage of a very economical and energy-efficient operation. In fact, energy and exergy efficiency are very important to offer reasonable and cheap charges for the airlines and the passengers using that airport, with a large profit remaining for the management. Therefore, operational economics is a major player besides several environmental issues. Figure 5.6 shows the electromechanical plant and the associated systems and equipment for an ideal airport.

5.4.3   Conventional Energy Systems and Equipment Only

In an airport with conventional energy systems, which is also referred to as Sample Airport 3, such a conceptual airport does not have any of the following electromechanical requirements for a truly sustainable airport:

- Cogeneration and trigeneration (absorption cooling during periods of low thermal loads)
- Heat pump technology with *COPEX* -->1 (Approaching one)
- Solar and wind energy (both subject to electromagnetic scrutiny by FAA and solar subject to glare effect scrutiny by FAA)

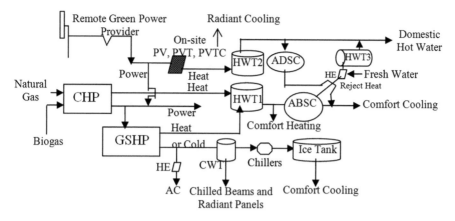

Fig. 5.6 An ideal airport electromechanical system

- Energy storage
- Energy from wastes
- Integrated waste management
- LowEx building

Under these design shortcomings and absence of all green systems and equipment, it will be very hard, if not impossible, to economically compete with other hub-candidate airports like Schiphol and Dubai, independent of the size and location of the airport. Some electromechanical design highlights for such an airport for a case study are given below:

- Boiler capacity (natural gas = 50 MWt)
- Chiller capacity = 40 MWe
- Electrical power source = grid with 120 MW capacity
- Power generators (per international guides)
- Heat pump technology = *none*
- Solar and wind energy (both subject to electromagnetic scrutiny by FAA and solar subject to glare effect scrutiny by FAA) = *none*
- Energy storage (both heat and cold) = *none*
- Energy from wastes (biogas, etc.) = *none*
- Integrated waste management = partial (water treatment)

Table 5.1 gives an estimate of annual fuel consumption and direct CO_2 emissions with preliminary assumptions of demand factors and base loads.

Instead, if a cogeneration plant for a 60% base power load is used with downsized boilers functioning only for thermal peaking and backup, the use of ground-source heat pumps with high *COP* value by employing LowEx building technologies, and some absorption coolers like in the case of Schiphol Airport, then the CO_2 emissions would decrease down to 280,000 ton/a from 1,040,000 ton/a. This means about 70% emissions reduction. Furthermore, natural gas savings would be around 45%.

Table 5.1 Estimated annual energy and environment performance transcript

Annual load	Energy demand, GW-h	CO_2 emissions[a]
Heat	230 GW-h	50,000 ton/a
Cold	100 GW-h (electric operated chillers)	110,000 ton/a
Electricity	800 GW-h (chiller loads included)	880,000 ton/a
		Total: 1,040,000 ton/a

[a]Excludes indirect emissions due to large exergy destructions in the boiler and chiller systems with grid power

5.5   Holistic Model for Sustainable Aviation

The diversity of the sustainability, safety, and hygiene parameters were brought to a common base by developing first a basic model and then expanding it to a nearly zero exergy airport model. Hygiene and human comfort were introduced into the model in terms of proper indexes, namely, hygiene index (HI) and human exergy loss index (HEXI). Airports are a breeding ground for potential worldwide epidemics with thousands of passengers traveling through each airport every day. Bomb attacks at Brussels Airport in the near past also point out the importance of security and safety. All these require additional energy, personnel, effort, and equipment, all of which add to the already high energy density of airports. Figure 5.7 shows the nexus of sustainable airports, which demand far beyond a simple LEED Green Airport Certificate. In this chapter, the first three elements of the nexus will be covered.

5.5.1   Energy

Equations 5.6-a through 5.6-c establish the energy and exergy metric of the nexus. For a green terminal building, the annual average of ψ_R value must be greater than 0.70.

$$\varepsilon_{sup} = \left(1 - \frac{T_{ref}}{T_{sup}}\right) \times (1\,kW\text{-}h) \quad \{\text{Unit exergy}\} \tag{5.6-a}$$

$$E_x = \varepsilon_{sup} \times Q_{sup} \quad \{\text{Energy, exergy}\} \tag{5.6-b}$$

$$\psi_R = 1 - \frac{\sum \varepsilon_{des}}{\varepsilon_{sup}} \quad \{\text{Exergy rationality}\} \tag{5.6-c}$$

Fig. 5.7 Four primary
nexus of a sustainable
airport

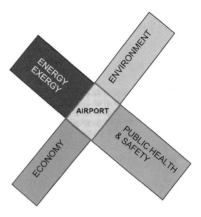

5.5.2   Environment

Equation 5.7-a, which is derived from the Rational Exergy Management Model
(REMM), establishes the environment metric of the nexus.

$$\Sigma CO_2 = \left[\frac{c_l}{\eta_l} + \frac{c_m}{\eta_m \eta_T}(1 - \psi_R)\right]Q_H + \frac{c_m}{\eta_m \eta_T}E \qquad (5.7\text{-a})$$

EDR is the ratio of emissions difference as given in Equation 5.7-b. CO_{2base} is
0.63 kg CO_2/kW-h as provided in Equation 5.7-c.

$$EDR = 1 - \left[\Sigma CO_2 / CO_{2base}\right] \qquad (5.7\text{-b})$$

$$CO_{2base} = \left[\frac{0.2}{0.85} + \frac{0.2}{0.35}(1 - 0.2)\right]0.5 + \frac{0.2}{0.35}0.5 = 0.63 \qquad (5.7\text{-c})$$

5.5.3   Economy

$$PES = \left[1 - \frac{1}{\left(\dfrac{CHPH\eta}{RefH\eta} + \dfrac{CHPE\eta}{RefE\eta}\right) \times \dfrac{(2 - Ref\psi_R)}{(2 - \psi_R)}}\right] \times 100 \qquad (5.8)$$

Equation 5.8 provides the economy metric, where *PES* must be greater than 35% in order to qualify for a green terminal status [12]. Based on these metrics of the nexus, three new sustainable airport definitions were made:

NZCB: A net-zero carbon terminal building, which on an annual basis, has an EDR value of one.

nZCB: A nearly-zero carbon terminal building, which, on an annual basis, has an EDR ≥ 0.80.

NZEXAP: It is an airport terminal complex, which has its own (local) district energy system, and it is connected to an external district energy system, which, on an annual basis, supplies the same total exergy of heat and power to the external district energy system (airport) as the total exergy of heat and power received.

On the other hand, a nearly zero exergy airport (*nZEXAP*) needs to have a district energy system on its own for energy and exergy rationality in the first place. The energy plant may be a centrally single one or more than one in the form of satellite plants. At any rate, the total exergy input from onsite renewable and waste energy sources to the airport complex that is served by the district energy system must be more than a minimum ratio of the total annual exergy demand of the airport complex. This ratio is 60% for summer and 70% for winter operation.

Sometimes, a new airport is constructed by sacrificing a large forest [13]. In this case, the amount of deforestation must be considered in the sustainability metric, and it must be kept at a minimum. This condition is given in Equation 5.9-a:

$$DFR = \frac{(DF \times s)/A_F}{(CO_{2base} \times E_T)/A_T} < 0.05 \tag{5.9-a}$$

Human exergy loss is especially important in terminal buildings during hectic transit times and overlay periods. Exergy loss is a much more important thermal comfort parameter, especially during transit times. Figure 5.8 shows the exergy comfort diagram developed by Shukuya et al. [14]. This figure shows that human exergy loss may be minimized for thermal comfort in terms of the mean radiant temperature (*MRT*) and the operative temperature (*OT*) rather than the simple air temperature (T_a). F_1 is the objective function for maximum comfort, which includes the indoor environmental quality indoor air quality (*IAQ*) factor. For the *OT* control in large buildings like terminal buildings, radiant panel heating, and cooling is very important not only for better human comfort reasons but also for reducing the sensible heating and cooling loads while demand temperatures are more moderate, which enhances heat pump performance coefficient largely.

The Suvarnabhumi Airport in Bangkok, Thailand, is home to the world's largest radiant panel cooling system. The developers covered an area of 150,000 m² with underfloor panels for the purpose of sensible cooling [15]. This system provides a maximum sensible cooling load of 70–80 W/m² at a floor temperature of 21 °C, while the dry-bulb (DB) indoor temperature is kept at 24 °C at the average human level (Fig. 5.9).

Fig. 5.8 Human body exergy comfort diagram [14]

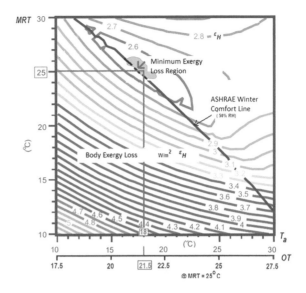

Fig. 5.9 Radiant panel cooling system in a green airport (LowEx) [15]

$$F_1 = 1/PMV' \quad \{Maximize\} \tag{5.9-b}$$

$$PMV' = PMV + a(0.6 - PR)^2 + b\varepsilon_H + c\left(\frac{IAQ_{ref}}{IAQ}\right) \quad \{Minimize\} \tag{5.9-c}$$

PMV is the predicted mean vote according to ASHRAE; *PMV'* is the expanded predicted mean vote according to the radiant to convective heat transfer split (*PR*) of the HVAC system(s) in the terminal. Coefficients *a*, *b*, and *c* are weighing factors mainly depending upon terminal size, annual passenger traffic, and climate.

Fig. 5.10 Basic model for nZEXAP

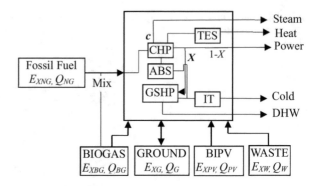

Referring to Fig. 5.7, the four nexus of a sustainable airport (SA), namely, energy and exergy, environment, public health and safety, and economy, a new model was developed, which is based on the second law of thermodynamics. This model is used to optimize the district energy plants and is shown in Fig. 5.10. This model allows considering combined heat and power (CHP) units, heat pumps, absorption and/or adsorption cooling units, thermal energy storage, and biogas. Solar and wind energy applications have strict constraints imposed by the US Federal Aviation Administration (FAA). Natural gas may be mixed with biogas, at any desired proportion. The objective is an effective decarbonization, which is described in detail by Soltero et al. [16], which may eventually lead to hundred-percent renewables [17]. Biogas exergy based on EXBG is the main parameter where only airport wastes are considered. According to another study [18], a cost-effective and technically feasible mix is about one-third. The mixing ratio determines EXNG (exergy of natural gas intake). The optimal CHP size for the now-known fuel intake can be found by considering the power-to-load ratio of demand, and usually CHP size is selected at the annual base load. Figure 5.10 shows that X is the split ratio of the power generated by the CHP system between heat pumps and other power demands of the airport complex. Other sustainable systems in the model like absorption chillers (ABS), ground-source heat pumps (GSHP), thermal energy storage (TES), and ice tank (IT) are solved from the first law.

Conventional HVAC systems like boilers and chillers are not subjected to optimization, because they have their own first law design procedures. The exergy terms shown in Fig. 5.10 include photovoltaic systems also. Waste heat from each building is used in that building; therefore, it is not subject to optimization either, but their contributions are subtracted from the district load.

5.5.4 Net- and near-Zero Definitions

An nZEXAP airport minimally receives 70% of the annual total exergy demand in winter. This ratio is 60% for winter because dominant cooling loads are more difficult to satisfy by renewables. If the annual average reaches 100%, nZEXAP becomes NZEXAP.

From the nZEXAP definition and Fig. 5.10, the first set of the two fundamental objectives of the optimization algorithm were derived.

$$OF_{1h} = \frac{E_{XBG} + E_{XG} + E_{XPV} + E_{XW}}{E_{XNG} + E_{XBG} + E_{XG} + E_{XPV} + E_{XW}} \geq 0.7, \quad \{\text{Winter}\} \qquad (5.10\text{-a})$$

$$OF_{1c} = \frac{E_{XBG} + E_{XG} + E_{XPV} + E_{XW}}{E_{XNG} + E_{XBG} + E_{XG} + E_{XPV} + E_{XW}} \geq 0.6. \quad \{\text{Summer}\} \qquad (5.10\text{-b})$$

Regarding a conventional airport, the objective functions which provide the nZEXAP index given in Equations 5.10-a and 5.10-b is just zero. For Schiphol Airport, this value is 0.72 from Equation 5.10-a at a mix ratio of six natural gases and one biogas.

There are not such exergy-based definitions yet in the literature. US Department of Energy defines only a zero-energy building in terms of the first law. In this definition, exergy differences between power, heat, and cold are not recognized. Actually, without doing so, avoidable (secondary) CO_2 emissions are imminent. Therefore:

- Unit exergy of heat, power, and cold must be distinguished.
- Exergy differences (destructions) must be factored into the net-zero or near-zero definitions like given above.
- These definitions must be based on seasonal operation instead of annual operations (annual average may be defined by some weighing functions).

These issues are addressed by new definitions for net or near building in terms of exergy (NZEXB, nZEXB). Biogas generation potential is, on the other hand, quite high in or around airports with thousands of passengers and hundreds of airplanes landing and taking off in a typical hub airport. District energy systems are a requirement in the built environment and rapidly expanding over the globe. 4DH systems not only distribute heat and power but at the same time distribute cold, service water, collect trash, and transmit data and communications. Regarding the Sample Airport 3, the nZEXAP index for both heating and cooling seasons (Equations 5.10-a and 5.10-b) is 0. For Schiphol Airport, which has a heating-dominated annual operation, the objective function from Equation 5.10-a is 0.72 at a natural gas-to-biogas ratio of 6:1. The value of 0.72 (>0.70) satisfies the green terminal requirement.

5.5.5   Exergy Balance in the Airport District

Exergy balance is a game changer in the net- or near-zero concept. When low-exergy HVAC is in consideration with low-exergy and intermittent sources of renewables, the game change becomes more serious. Figure 5.5 shows a case study where the building delivers 30 °C water but receives 40 °C warmer, thus with higher exergy in the same amount. Then, a deficit in exergy occurs. A similar deficit is present for cooling. The ideal (net-zero) conditions are given in Equations 5.11 and 5.12. Electricity exchange quality is assumed identical.

$$E_{xsuph} = \varepsilon_{suph} \times Q_{suph} = E_{xreth} = \varepsilon_{reth} \times Q_{reth} \quad \{\text{Ideal balance in heating}\} \quad (5.11)$$

$$E_{xsupc} = \varepsilon_{supc} \times Q_{supc} = E_{xretc} = \varepsilon_{retc} \times Q_{retc} \quad \{\text{Ideal balance in cooling}\} \quad (5.12)$$

5.6 Design and Rating and Metrics

Existing metrics in the literature have been based on the first law of thermodynamics. As mentioned in the earlier sections of this chapter, the exergy balance of the exergy exchange among supply and demand points are very important, and usually, this kind of analysis is a game changer. In order to progress in this direction, new metrics based on exergy are developed. The new metrics in a condensed form are given in Table 5.2.

5.7 Conclusion

In this chapter, the importance of exergy rationality in net-zero or near-zero airports connected to the 4DE systems of the neighboring cities was discussed, and new metrics were defined. It has been shown that net-zero or near-zero exergy definitions along with new emissions metrics surpass the green terminal certification programs in many aspects. A green terminal by the certificate may not be actually green. For example, if an airport receives an amount of X MtoEX of the annual composite total of exergy (power, heat, cold, etc.) from a DE system and annually supplies an amount of Y MtoEX to the DE system, and if $X > Y$, then there is an exergy deficit, and the airport may not be labeled to be truly sustainable, because it causes additional exergy destruction and CO_2 emissions. These additional items must be considered in a more insightful sustainability analysis of an airport, which was addressed by the new metrics presented in this chapter. Furthermore, in the quest for greener airports, it is hereby suggested that future studies must also consider the airside of an airport. For example, the location of an airport must be carefully selected in terms of minimizing the fuel spending of the aircraft at the approach and takeoff paths. None of the current green terminal certification programs consider the airside of the airport and consider only the landside of the terminal building. A holistic approach must go well beyond the terminal building itself, including the HVAC services given to docked aircraft to the gate.

In addition to the current optimization algorithm, some more optimization task awaits at the building side of an nZEXAP airport. Especially for new airports, or deep renovation projects of terminal projects, there exists a chance to further opti-

Table 5.2 Green airport rating metrics

Metric		Definition	Condition
ψ_R		REMM efficiency	>0.70
η		First law efficiency	>0.75
η_{II}		Second law efficiency	>0.50
PES_R		Exergy-based-primary energy savings ratio	50%
PER		Primary energy ratio	>1.2
SF^a	Heating	Load shaving factor	0.3
	Cooling		0.4
DF		Diversity factor	0.2
EDR		Emissions difference ratio	>0.80
LFR		Load fluctuation ratio	<0.20
$MtoEX_s/$ $MtoEX_r$		Annual composite exergy exchange ratio between the airport and DE	>0.80
AER		Alternative energy ratio	>0.50
$COPEX$		Exergy-based COP	approach to 1
MRE		Moisture removal efficiency	>4.5
DFR		Deforestation ratio	<0.05
HI		Hygiene index	≥80/100
$NZEXAP$		Nearly zero exergy airport	≥0.70 in winter ≥60 in summer
IAQ/IAQ_{ref}		Indoor air quality index	≥90/100
ε_H		Exergy comfort index	≤2.5 W/m²
PMV'		Expanded PMV index	≥1.5 PMV

[a]With energy storage

mize the heating and cooling heat transfer optimization: lower the heating fluid temperature demand or higher the cooling temperature demand from the GSHP, COP (coefficient of performance) of the heat pump increases. This brings LowEx (low-exergy) buildings forward. However, higher or lower the fluid temperature in cooling or heating, respectively, the temperature drop between supply and return fluid temperature needs to be smaller, like in chilled beams versus fan coils, such that pumping size and cost increase and usually heating or cooling equipment must be oversized. These costs factors and exergy inputs and outputs to and from GSHP, pumps, and oversizing establish a challenging optimization problem. But the result will be rewarding for the economy, energy savings, and the environment.

Symbols

A_F	Deforested total area, m²
A_T	Total ground area of all terminal buildings, m²
AER	Alternative energy ratio to the total energy demand
CHPHη	First law efficiency of heat generation (by CHP)
CHPHη	First law efficiency of power generation (by CHP)

CO_2	Carbon dioxide emission, kg CO_2
CO_{2base}	0.63 kg CO_2/kW-h
COP	Coefficient of performance according to the first law
COPEX	Coefficient of performance according to the second law
DF	Number of deforested mature trees
DFR	Deforestation ratio
E	Electric power (energy) demand, kW or kW-h
E_T	Total annual energy demand ($Q_H + E$), kW-h
E_x	Exergy, kW or kW-h
EDR	CO_2 emissions ratio
F_1	Objective function for human thermal comfort
F_2	Cost minimization objective function
HI	Hygiene index
IAQ	Indoor air quality
m	Ratio of natural gas to biogas mix for CHP fuel intake
MRT	Mean radiant temperature
Mtoe	Megaton of oil equivalent
MtoEX	Megaton of oil exergy equivalent
OT	Operative temperature
PES	Primary energy savings percentage
P_F	Fan (or pump) power, kW
PMV	Predicted mean vote according to ASHRAE
PMV'	Expanded PMV
PR	Radiant to convective heat transfer split in HVAC
Q_H	Thermal (heat or cold) power (energy) demand, kW-h, or kW
RefEη	Reference value for partial power generation efficiency of CHP
RefHη	Reference value for partial thermal generation efficiency of CHP
s	Annual CO_2 sequestration capacity of a tree, kgCO_2/a
T	Temperature, K
T_o	Outdoor air temperature or heat pump return temperature, K
T_r	Preheated air temperature by energy recovery or heat pump supply temperature, K
T_{ref}	Reference environment temperature, K
X	Power split ratio between GSHP and other services at the airport

Greek Symbols

ψ_R	Rational Exergy Management Model (REMM) efficiency
Refψ_R	Reference value of ψ_R, 0.3
η	First law efficiency
η_{EX}	Second law efficiency
η_T	Power transmission, distribution efficiency
η_m	Power plant efficiency
ε_{dem}	Unit demand exergy, kW/kW, or kW-h/kW-h
ε_{des}	Unit destroyed exergy, kW/kW, kW-h/kW-h
ε_{sup}	Unit supplied exergy, kW/kW, or kW-h/kW-h

Subscripts

dem	Demand
E	Electric
H	Thermal
l, m	Local power plant, distant power plant
ref	Reference
sup	Supply
r	Received from DE system
s	Supplied to DE system

Acronyms

ABS	Absorption chiller
ADS	Adsorption chiller
ASHRAE	American Society of Heating, Refrigerating, and Air-Conditioning Engineers Inc.
ATES	Aquifer thermal energy storage
CPH	Copenhagen Airport
CHP	Combined heat and power
CWT	Cold-water tank
DB	Dry bulb (temperature)
DE	District energy system
4DH	Fourth-generation district heating system
DH	District heating system
DOE	United States Department of Energy
FAA	Federal Aviation Administration
FT	Fischer-Tropsch process
GSHP	Ground-source heat pump
HE	Heat exchanger
HEXI	Human exergy loss index
HI	Hygiene index
HVAC	Heating, ventilating, and air conditioning
HWT	Hot-water tank
LEED	Leadership in Energy and Environmental Design
LowEx	Low exergy (building)
MRO	Maintenance, repair, and overhaul
nZCA	Nearly zero carbon aviation
nZEXA	Nearly zero exergy aviation
NZEAP	Net-zero energy airport
IPCC	Intergovernmental Panel on Climate Change
PV	Photovoltaic panel
PVT	Photovoltaic and thermal panel
PVTC	Photovoltaic-thermal and cooling panel

SWWCP	Solar, wave, wind energy combined offshore platform
IT	Ice tank
SA	Sustainable aviation
TES	Thermal energy storage
UN	United Nations

References

1. ATAG, Air Transport and Aviation Group. Aviation and climate change. https://aviationbenefits.org/environmental-efficiency/our-climate-plan/aviation-and-climate-change/. Accessed 05 July 2018
2. Kılkış Ş, Kılkış Ş (2017) Benchmarking aircraft metabolism based on a sustainable airline index. J Clean Prod 167, 1068–1083, https://doi.org/10.1016/j.jclepro.2017.03.183
3. Kılkış B, Kılkış Ş, Kılkış Ş (2019) A simplistic flight model for exergy embodiment of composite materials towards nearly-zero exergy aviation, Int. J. Sustainable Aviation, in print
4. Kılkış B, Kılkış, Ş (2018) An electric aircraft but not exactly, ISSA 2018 conference paper, Kiev, Ukraine
5. Kilkis B, Kilkis, Ş, Kilkis, Ş (2019) A novel off-shore wind turbine platform over H_2S reserves in the black sea that combines solar and wave energy to generate hydrogen fuel for a hydrogen-economy city and jet fuel with local coal-based power plant CO_2 sequestration. *Energies*, Under Review
6. Rich A, Patel J (2015) Carbon disulfide (CS_2) mechanisms in formation of atmospheric carbon dioxide (CO_2) formation from unconventional shale gas extraction and processing operations and global climate change. Environ Health Insights 9(Suppl 1):35–39. https://www.ncbi.nlm.nih.gov/pubmed/25987843. Accessed 13 Aug 2018
7. European Commission (EC), Fuel produced from sunlight, CO_2 and water: an alternative for jet fuel? Science for Environment Policy, 9 Sept 2016. Issue 469. http://ec.europa.eu/environment/integration/research/newsalert/pdf/fuel_sunlight_co2_water_alternative_jet_fuel_469na1_en.pdf. Accessed 13 Aug 2018
8. Kılkış Ş (2012) A net-zero building application and its role in exergy-aware local energy strategies for sustainability. Energy Convers Manag 63:208–217, https://doi.org/10.1016/j.enconman.2012.02.029
9. Kilkis Ş (2014) Energy system analysis of a pilot net-zero exergy district. Energy Convers Manag 87:1077–1092, https://doi.org/10.1016/j.enconman.2014.05.014
10. Kılkış, B, Kılkış, Ş (2018) Hydrogen Economy Model for Nearly Net-Zero Cities with Exergy Rationale and Energy-Water Nexus, Energies 11(5), 1226; https://doi.org/10.3390/en11051226
11. Kılkış Ş, Kılkış Ş (2015) Benchmarking airports based on a sustainability ranking index. J Clean Prod 130:248–259, https://doi.org/10.1016/j.jclepro.2015.09.031
12. Kılkış B, Kılkış Ş (2015) Cogeneration with renewable energy sources, TTMD publication No 32. Doğa Pub. Co., İstanbul. ISBN:978-975-6263-25-9, 371 pp (*in Turkish*)
13. Kılkış B (2014) Energy consumption and CO_2 emissions responsibility of airport terminal buildings: a case study for the future Istanbul airport. Energ Buildings 76:109–118, https://doi.org/10.1016/j.enbuild.2014.02.049
14. IEA Annex 37-LowEx. 2002. (2002) IEA Annex 37 News No: 5-human body exergy consumption and thermal comfort, IEA Annex 37-LowEx.
15. Cubick R (2016) Bangkok airport: the world's largest radiant cooling system. https://www.uponor.hk/radiant-cooling-blog/bangkok-airport-the-worlds-largest-radiant-cooling-system/. Accessed 1 Oct 2018

16. Soltero VM, Chacartegui R, Ortiz C, Velázquez R (2016) Evaluation of the potential of natural gas district heating cogeneration in Spain as a tool for decarbonization of the economy. Energy 115(3):1513–1532.
17. Markovska N, et al. (2016) Addressing the main challenges of energy security in the twenty-first century. Contributions of the conferences on sustainable development of energy, water, and environment systems. Energy 115(3):1504–1512.
18. Taşeli B, Kılkış B (2016) Ecological sanitation, organic animal farm, and cogeneration: closing the loop in achieving sustainable development–a concept study with on-site biogas fueled trigeneration retrofit in a 900 bed university hospital. Energ Buildings 129:102–119.

Part III
Aircraft

This part consists of two chapters. They are "A Short Brief on the Aircraft History and Anatomy" and "Sustainable Aircraft Design."

In Chap. 6, firstly, aircraft history is expressed and then aircraft anatomy (only fuselage, wing, empennage, and landing gear) is explained.

In Chap. 7, sustainable aircraft design is explicated with different literature. Authors also clarify aircraft design methodologies with its phases in this chapter.

Chapter 6
A Short Brief on the Aircraft History and Anatomy

Murat Ayar and K. Melih Guleren

The history of aviation and aircraft anatomy are two strongly related topics that lead the researchers in aviation to produce valuable studies should be known in the understanding of aviation-related studies. In this section, firstly the known history of aviation will be briefly given. It is generally focused on the people and the events. Finally, future technologies and hot topics are mentioned. In the other section, the basic components of an aeroplane are explained. These components are grouped as main and sub-systems. These parts are explained in a simple way as their functions and structural features.

6.1 Introduction

As technology develops, aviation is developing, and technology is developing in line with the needs of aviation. For this reason, aviation and aviation technologies are very important areas. In order to understand aviation and technology, it is necessary to know the basics of the related subjects. Aviation history and aircraft anatomy are on the basis of aviation. In this section, these subjects are explained at a basic level. Only the power plant section is not mentioned, because this section is described in the second volume of this book.

The history of aviation should be handled by considering the important stages and the dedicated labour. A detailed understanding of the successful and failed aircraft designs will help the innovative future studies.

The history of aviation began with tower jumpers and lightweight balloons. Distances has reduced with the birth of aircraft. There has been many remarkable people in aviation history. Actually, the aviation started with the personal efforts

M. Ayar · K. M. Guleren (✉)
Eskisehir Technical University, Faculty of Aeronautics and Astronautics, Eskisehir, Turkey
e-mail: muratayar@eskisehir.edu.tr; kmguleren@eskisehir.edu.tr

© Springer Nature Switzerland AG 2019
T. H. Karakoc et al. (eds.), *Sustainable Aviation*,
https://doi.org/10.1007/978-3-030-14195-0_6

sown by these people. Today is getting more institutionalized, there are many national and international institutions and componies working in aviation.

Aircraft components and their functionality should be understood well. When examining aircraft anatomy, it is necessary to pay attention to the missions of the components as well as the types of materials. Although there are many aircraft types, the general aircraft parts are the same, and they perform the same mission. Depending on the size of the aircraft, the tasks do not change, but the structure types may change.

In general, the aircraft can be classified in several ways. The first is the classification according to the type of operation. This classification comprises three types: general aviation, commercial and military. General aviation aircraft range from gliders and crop dusters to rotorcraft and corporate business jets. Commercial aeroplanes are used on scheduled flights by airline companies. Finally, military aircraft are used for military purposes by the armies of countries. No matter which class, aircraft anatomy is almost the same for all classes or has very small differences.

6.2   Aircraft History

6.2.1   Timeline of Aircraft

400 B.C.	Chinese invented the kites.
875	Ibn Firnas made the first plane and flight.
1485	Leonardo da Vinci designed a wing-flapping aircraft.
1783	Montgolfier brothers launched the first hot air balloon with animals in it.
1849	George Cayley designed the first glider that can carry a person.
1891	Otto Lilienthal designed and flew the first glider for long flights.

1903	Wright brothers invented the first controllable motor-powered aeroplane.
1927	Charles Lindbergh flew across the Atlantic Ocean without any stop.
1947	Chuck Yeager broke the sound barrier for the first time.
1986	Dick Rutan and Jeana Yeager flew around the world without any stop.
1997	The NASA's Pathfinder, the solar-powered aircraft, flew above the troposphere.

6.2.2   Pioneers of the Aircraft (Before the 1700s)

The first experimenters included Heron, the inventor of Alexandria (around 50 A.D.), worked together on the steam-powered aeolipile. Around 1500 A.D., Leonardo da Vinci drafted a chimney jack that used hot gases flowing from a chimney to run fan-like blades. The principles of spit and aeolipile were first proposed by Isaac Newton, who formulated the law of motion in 1687.

Abbas Ibn Firnas was a Muslim astronomist who lived in the ninth century. He was able to manufacture the first plane and fly with it. Historical sources indicate that the Islamic scholar Ibn Firnas, after very long tedious studies, invented a methodology based on installing fabric on top the wing to decrease the air resistance which yields much better flight. In 875, Ibn Firnas built a glider and left himself out of the tower like a bird. It was conveyed that he had been gliding like a bird for a long time and then slowly descended to the ground. The flight was largely successful and was watched by a crowded group of people. But his landing was a little rough, hurting his bones. Some were told that he did not take adequate measures on the tail of the glider and did not design the tail enough for better manoeuvres [1].

The concept of "flying machine" was first proposed by Roger Bacon in the thirteenth century. Bacon was regarded as one of the founders of modern scientific tradition. Bacon predicted that people could fly with a machine that has flapping wings. The ornithopter was the name given to flying vehicles, which mimics the principle of flapping wings of the birds.

This mysterious genius of Renaissance, Leonardo da Vinci, was born in 1452 in a simple house very close to the village of Vinci in Italy. da Vinci was one of the famous scientists of the Renaissance. Although he was known as a painter, da Vinci was also an anatomist and an engineer. Leonardo fancied man escaping centrism and examined the flight and the anatomy of birds by resorting to nature [2]. His first studies on flying started in the 1480s. In his early work, he designed mechanisms for the reciprocal or cyclic motion transformation in the mechanical motion of various wing parts. In 1487, he drew the first design. He developed a flight machine by examining a dragonfly. At the end of the fifteenth century, he designed a moving wing that could lift an object weighing about 90 kg and then designed a machine that had a propeller-like screw, which was considered the pioneer of the modern helicopter. In later years, he designed wings with a hand lever to enable movement [3].

Ahmed Celebi was the first Turkish to fly in the world. It is assumed that he lived in the seventeenth century, during the reign of Sultan Murad IV. Celebi was known as Hezarfen in public because of his extensive knowledge. He conducted experiments in Okmeydani in order to examine the endurance of the wings which encouraged him to do his historic flight. One morning, in front of the people gathered in the coast, he released himself from the tower of Galata, moving the wings over the Bosphorus and went down to the Uskudar district. Celebi flew in this way and crossed the Bosphorus about 3500 m [4].

6.2.3 Industrial Age Flights (1700–1900)

The first manned flight took place in Paris in 1783. Jean-François Pilâtre de Rozier and Francois d'Arlandes have travelled 8 km using a hot air balloon invented by the Montgolfier brothers. The balloon was heated by the wood fire and could not be controlled, which meant that the wind was flying wherever it took it [5].

The first Zeppelin was 128 m long and 11 m in diameter. Its aluminium skeleton was covered with a cotton cloth. The skeleton contained hydrogen gas bubbles. The airship aired on 2 July 1900, flying from a height of 400 m, and took a 6-km road in 17 minutes and 30 seconds [6].

Otto Lilienthal, a German inventor, is known as the the pioneer of gliding. He wore the flapping wings for an unusual flight attempt in 1868. Later he examined the flight of the birds, especially the storks. He thought that horizontal flights could be made with the help of air currents without having to push the wind. In 1877, he attempted to a cruise flight with the help of wings. His real success came in 1891 with a winged flight vehicle [7]. Lilienthal was able to move and control successfully a glider which was made of wheels and stretched rags of 20 kg weight [8].

The pioneer of aerodynamics is accepted globally as George Cayley. Although he was unable to find enough financial means, Cayley spent a significant part of his life studying the basic principles of flight. Cayley's work on aerodynamics includes the determination of lifting and thrust forces, automatic balancing in three coordinates, reducing air drag and construction and testing of single-, double- and triple-wing aircraft and winged missiles [9]. While studying the air on the wings, he realised that the wings produce actually the lift force and the thrust. Therefore, there should be something another mechanism to create the thrust force. As he saw the benefits of a fixed wing in aeroplane design, he clearly pointed out the function of the aircraft tail, lifting surfaces, propeller and rudder in flight. Thus, he outlined the modern aircraft [10].

6.2.4   Born of Aircraft

Wright brothers completed the construction of the first motor aircraft, Flyer 1, in 1903. On the morning of 17 December 1903, Orville Wright launched the first manned motorized and controlled flight of aviation history from a flat surface with no take-off support, except for the engine of the aircraft [11].

The Wright brothers made more than 80 public flights by calling their friends, neighbours and journalists during 1904 and 1905. However, they showed little interest in these invitations.

On 13 September 1906, Alberto Santos-Dumont made a public flight in Europe. He flew a distance of 221 meters by means a chambered wing and an elevator. Since this aircraft does not require a catapult for any counter-wind and take-off, some people regarded the first flight as an engine-powered flight [12].

Henry Farman and John William Dunne were two British inventors working on separate engines. In January 1908, Farman won the Grand Prix Aviation Award for a machine that was flying longer than 1 km [13]. On 14 May 1908, the Wright brothers took their flight to Charlie Furnas, a flight that would be considered the first two-person flight.

Aeroplanes were also included in the military service immediately they were met. Bulgaria was the first country to use aircraft for military purposes. They were used in the First Balkan War (1912–1913) to explore the Ottoman fronts. The World War I was the first war for which the aircraft was seen to perform the missions of attack, defence and exploration [14].

Commercial aviation, after the World War II, began to evolve through goods transport.

6.2.5   Jet Aircrafts

In the 1930s and 1940s, there were many pioneers in jet engines like an English engineer Frank Whittle and a German engineer Hans von Ohain. They were working without knowing each other's work on the jet engine. Sir Frank Whittle began to work first in this field, but Hans von Ohain was the inventor who had previously completed the jet engine.

In his early years of the 1930s, Hans von Ohain began to work on a jet engine when he was studying in Germany. During the year 1935, he continued his work on the test engine and applied his thoughts. Upon getting positive results from the engine, he told his studies to aircraft manufacturer Ernst Heinkel and asked for his support. Watching the work of Hans von Ohain, Ernst Heinkel produced the world's fastest aircraft in a very short time. S-1 turbojet engine was tested with hydrogen fuel in 1937, and 250 pounds of propulsion power was provided from the engine [15].

Frank Whittle began his jet engine work in 1928 while serving as a British Air Force officer. In 1928, Whittle published an article on systems such as gas turbines or jet propulsion, instead of internal combustion engines that allowed aeroplanes to fly. He filed a patent application for the first turbojet engine he invented in January 1928. His application was approved in 1931. Its design was adopted by many people. However, the Air Ministry and the aviation industry showed little interest in this design. Thanks to de Whittle, in 1941 the Gloster-Whittle E28/39 aircraft with Whittle's jet engine aired [16].

De Havilland Comet was considered as the world's first jet engine manufactured by the British De Havilland Aircraft Company. It was launched in 1949 and made the first commercial flight in May 1952 [17].

In 1961, the sky was no longer a limit for manned flight. Yuri Gagarin left the earth and made a 108-minute orbit flight. This phase accelerated the space race, which began in 1957 when Sputnik was launched by the Soviet Union into space.

6.2.6 World War II Era and After

During World War II, the British introduced the new radar technology Spitfire fighter. Dozens of Japanese warriors and bombers launched from aircraft carriers destroyed the US Pacific Fleet at Pearl Harbor in 1941, while Americans launched large-scale bombers such as the B-17 de Flying Castle 19 and B-29. The bomber was dropping the atomic bomb in Japan. On the seafront, aircraft carriers were the largest ships used in battle; a steam device was used to push the plane for departure, and a hook at the rear of the plane was used to catch a cable in the lane and stop it after the descent.

On 14 October 1947, a sound, which was never heard before, was heard in the Mojave Desert: the sonic explosion of a plane reaching Mach 1. The plane was the Bell X-1 and Chuck Yeager in the controls [18].

The USA immediately started to design a new project immediately after World War II. The USA needed a long-range subsonic strategic bombardment capable of giving orders to an enemy position deep in the airspace or a huge amount of conventional or nuclear orders. The Boeing B-52 Stratofortress was used in the Vietnam War. It was so powerful and also useful that it remained in service for a very long time. With a crew of five, the B-52 was powered by eight Pratt & Whitney TF33 turbofan engines, each of which has 17,000 lb thrust force. The net result reached a speed of 650 mph with a maximum range of 9000 miles without refuelling [19].

This was probably the most advanced aircraft ever developed until 1960s and was definitely one of the highest performing military aircraft of all time. The operational history of Lockheed's SR-71 Blackbird began with the Vietnam War. Designed to work with a two-man crew, the SR-71 was powered by twin Pratt & Whitney J58-1 continuous overflow turbojet engines. Although the numbers were classified, the SR-71 had a minimum Mach 3.3 speed. The range was estimated as 2900 nautical miles with a service height of 85,000 feet [17]. A total of 32 aircraft were built.

While NASA continues its aviation activities, it increases the R&D studies in order to yield more efficient, lower emission and noiseless operation of aircraft. In addition to various R&D efforts, NASA is organizing competitions to create more technological activities for the future.

When we look at the current trends of aviation technologies, it is seen that some of the issues are studied worldwide. These are sustainable aviation, electric airplanes, supersonic aircraft, biofuel airplanes, flying cars, flying men and unmanned aerial vehicles (UAV). These six topics constitute the future main targets of the countries and authorities.

6.3   Aircraft Anatomy

In order to fly safely, it is necessary to have an adequate structural strength for the aircraft which has to be maintained according to the rules dictated by the aviation authority of the aircraft. In order to satisfy the requirements of their defined performance, it is a must for the aircraft manufacturer to provide the materials and design of the aircraft structures with sufficient strength. The aircraft can be classified into five main parts: fuselage, wing, empennage, power plant and landing gear.

6.3.1   Fuselage

The fuselage is one of the main parts of the airframe designed to accommodate crew, passengers and cargo. It has many kinds of design and size according to the aircraft mission. In a jet fighter, for example, the fuselage is mainly composed of a cockpit which is long enough for the pilots and avionics, but a commercial passenger aircraft has a larger cockpit and a very long cabin with separate decks for passengers and cargo.

The aircraft fuselage seen from the can be divided into two parts. These are called as the upper lobe (upper half slice) and the lower lobe (lower half slice). These half slices combine at approximately floor level, which separates the passenger and cargo compartments from each other. The upper part of the floor structure carries the passenger and flight compartments. The cabin is closed by the pressure bulkhead from the front and rear end. These bulkheads distinguish between the outside air pressure and the ambient pressure. In this region, the cladding is permanently covered.

The cargo compartment is closed by the lower half slice below the cabin floor. This area of the aircraft is basically composed of two parts and is divided by various elements. These elements are nose landing gear wheel wells, centre wing box and main landing gear wheel wells.

The structure of the floor behind the rear pressure chamber does not continue, which is reserved for vertical fin, horizontal stabilizer connections and APU compartment.

The loads acting on the aircraft are changed during various phases of the flight, including taxing, take-off, landing, cruise and even pressurization when required. Basically, the airframe resembles a hollow tube supported by beams from its wings. As a result of this, the loads caused by various acts such as sudden manoeuvres affect the entire body. The central wing box and the main landing gear housing areas, which are important parts of the aircraft, are combined with special design features to form a continuous body skeletal structure. In the region of the wing connections, keel beam is used to ensure the continuity of the skeleton.

In the centre wing part, two beams with six flanges are used generally for the wing body connections. The body trim is attached to the top of these flanges. The other flanges connect to the centre wing box. Main landing gear and wing weights are transmitted to the body via the landing gear support beams in this section. The openings caused by all the doors and windows on the body skin are specially reinforced. These openings in the coating are the areas where some of the loads are acted.

The body consists of four separate parts. After they are manufactured, these parts are assembled in one place. The front three parts are the cockpit, flight and cargo compartments.

The nose fuselage section has a flight compartment (cockpit), front entrance door and front service door. Under the cabin floor, there is a nose landing gear housing, which is accessed by the outer door and the ladder, where it hosts an electronic compartment. The upper side of the frontal fuselage is occupied by the forward region of the pressure cabin, whereas the lower side is reserved for the cargo section. The aft fuselage can be considered between the end of the frontal fuselage and the aft pressure chamber. On the upper lobe, there are rear passenger compartment, emergency exits, rear entrance door and rear service door. There is a central wing box on the lower lobe, a rear landing compartment box housing and a rear cargo compartment. Tail fuselage section starts from the rear pressure wall. Generally, there is a vertical stabilizer attached to the aircraft perpendicular to the horizontal stabilizer. The tail cone is located at the back of the tail. The bottom of the horizontal stabilizer is separated by fire compartments. This is the APU compartment. The horizontal stabilizer is hinged to the aircraft structure via its own cage structure. Stabilizer connections are connected to the front and rear spar (longitudinally extending beams).

6.3.1.1 Fuselage Types

There are three types of the main fuselage or general structure of the aircraft. These are truss, monocoque and semi-monocoque types. The primary structure is the main structure for the elements that carry the aircraft, whereas the secondary structure is the ones except the main structure of the fuselage and auxiliary structures.

6.3.1.1.1 Truss-Type Structure

In this type of structure, the aircraft frame is generally formed by steel tubes which are welded to each other diagonally. In this method, the structure of the aircraft can be produced as aluminium pipes when lightness is desired.

6.3.1.1.2 Monocoque-Type Structure

Monocoque-type structure consists of a body, a moulded frame, pressure chambers and a separately bonded coating sheet. The strength of the body is related to the properties of the coating material. In monocoque structures, longitudinally extending structural elements such as stringer and longeron are not used, and the body plate is connected directly to the frames. This type of structure method is difficult for the body sheet to wrap the structure and to provide a stable weight for flight.

6.3.1.1.3 Semi-Monocoque-Type Structure

The semi-monocoque-type structure is the most common structure applied in today's aircraft. In this structured method, structural elements such as a bulkhead, frame-like rib elements and longeron and keel beam are used. Stringers prevent the torsion of the body as longitudinal support to the coating sheet. Semi-monocoque fuselages are obtained by manufacturing a plurality of parts and subsequently joining them together.

In modern passenger planes, generally, semi-monocular structure method is applied. Manufacturing and maintenance are easy with this type of structure. Moreover, the aircraft service life is extended because the loads on the aircraft are distributed to the various components. Advanced passenger planes consist essentially of longitudinally extending strippers and longerons and frames that support the covering sheet.

6.3.1.2 Fuselage Components

6.3.1.2.1 Frames

Frames are the main building elements forming the shape of the fuselage. They shorten the length of the stringer and prevent structural imbalance. They are responsible to carry the shear and tensile loads on fuselage structure.

6.3.1.2.2 Stringers

Stringers are the elements that connect the frame structure elements to each other and to the fuselage sheet. They are placed also inside the wings to connect the wings. They prevent the wings and the fuselage form buckling and breaking.

They are designed to carry loads due to bending and shearing stresses and cabin pressure induced from the fuselage structure. The stringers are connected to the rivets through the frames. They are in the form of angular-shaped or "T"-shaped clips. Clips are located on the inner surface of the body plate depending on the stringer and the frame. Their purpose is to transfer the pressure load from the body sheet to the frames.

6.3.1.2.3 Longerons

Longerons are parts of the aircraft structure designed to enhance rigidity and strength of the frame. Longerons are like stringers but they are much longer than stringers. Longerons attach directly to the frame of the aircraft using bolts.

6.3.1.2.4 Bulkheads

Bulkheads are similar to the frames but much stronger. They are located in certain places on the fuselage. They carry and distribute different loads on the body. They are located on the surface of the landing gear and tail section in the area of the wing connections on the body. These regions are the parts with high-intensity forces on the fuselage. They meet the major stresses on the structure. In addition to the front and rear of the body, pressure bulkheads are installed on different structures. Front and rear pressure bulkheads are wall-like structures. The fuselage structure between the front and rear pressure bulkheads balances the pressurized cabin and the outside.

6.3.1.2.5 Doublers

Doublers are reinforcement layers used around window and door openings where reinforcement is required. At the same time, repairing surfaces (fuselage patches) used in repaired surfaces are also called doublers.

6.3.1.2.6 Fuselage Skin

The fuselage skin forms the outer shape of the body. Fuselage sheet and reinforcement structures are the longest units of the fuselage. It straps the structure as the stringer and frames are aligned at regular intervals and also carries the main loads. The body surface is a structure that helps to prevent buckling and wrinkling on the structure. The fuselage skin consists of coated aluminium, anodized aluminium with chromic acid or iodine-treated aluminium to prevent corrosion.

6.3.1.2.7 Floor Beams

Flooring beams are the structures that are installed on the cabin floor. Usually, they form the structure that separates the cabin and cargo sections. They are horizontal structures extending to the fuselage skin. They carry pressure loads by connecting to the frames. They stand to the pressurization of the cabin. Seat tracks are the structures that carry the normal loads of floor panels.

There are skin shear ties on both sides of the hull, which are raised by the floor beam and lead to the distribution of pressure loads on the cabin.

6.3.1.2.8 Keel Beams

Keel beams are located in the centre wing of the body. They support the landing gear well where maximum bending occurs. They form also the possible structure for the landing gear to be collected in its housing. They are the largest beams in the fuselage structure.

6.3.2 Wing

Wings provide the lifting force which keeps the aircraft in the air. Since wings have the most suitable place among the other parts of the aircraft, they carry most of the fuel. The surfaces that change the aerodynamic forces; such as slates, flaps, ailerons and spoilers, are located on the wings. Since the wings are subjected to most of the aerodynamic forces, the maintenance intervals are quite short; frequent dismantling and assembly of these parts are necessary to perform repairs and checks.

The wing and the empennage may encounter deformations such as cracking and breaking due to the effects of forces and moments acting on them.

6.3.2.1 Wing Structure Types

Wing types according to their structure are classified as monospar, multisport and box beam. Monospar wings are not found to be of much use. However, with the support of a structure in the shape of L or T and other beams that can be added to monospar wings, these wings can also be used in some aircraft.

Multisport wing application, with more than one carrier beam, extends in the wing section. This type of wings is equipped with ribs connecting each spar and independent pressure chambers.

The box beam type, which has a common application area, is equipped with ribs-formed partitions between two spars. The inner surface of the wing is supported by stringer or stiffeners. In the advanced box beam wing manufacturing, the wing skin is produced with chip machining techniques together with stiffeners on it.

6.3.2.2 Wing Components

Wings are designed considering many factors, including aircraft weight, speed, range, endurance and a specific mission. Therefore, there is a different wing structure and shape for each aircraft type. Although the wing shapes are different, the basic structural elements are the same.

6.3.2.2.1 Spars

Spars are the basic load-bearing elements forming the interior structure. According to the aircraft type, one, two or three pieces can be found in the wing structure. Spar production varies also according to aircraft type. Small aircraft might be made of carbon composites or aluminium, while larger aircraft might use steel alloy advanced materials. Spars are connected to the ribs.

6.3.2.2.2 Ribs

The ribs are installed according to the cross section of the wings. The ribs are also connected to the stringers along the wings. Load transfer takes place from the surface coating to the stringer.

6.3.2.3 Wing Control Surfaces

Main wing control surfaces are aileron, flaps-slats, spoilers and tabs. The structure and dimensions of these control surfaces vary according to the aircraft model and dimensions.

6.3.2.3.1 Ailerons

Ailerons are used to roll the aircraft when required. The ailerons are used to bank the aircraft; it causes one wing to rise up and the other wing to move down. The ailerons work by changing the effective shape of the airfoil close to the tip of the wing. Changing the angle of deflection of wing tips will change the amount of lift generated by the wing. Ailerons often work in opposition to each other like when the right one is deflected upwards, the left is deflected downwards and vice versa.

Ailerons are composed of beams on the leading and trailing edges and the structure between these beams. Ailerons are placed with the help of balance panels located between the leading edge and the rear spar. Each bearing of balance panel is accessible with removable cover panels. Balance tab on the trailing edge is available.

6.3.2.3.2   Flaps and Slats

Aircraft are required to increase the effective chamber of the wing during take-off and landing to provide enough lift to perform the mission. It is done by placing moving parts on the leading and trailing edges of the wings. The moving parts at the leading edge are called slats, while those at the trailing edge are called flaps. Flaps and slats move through metal parts placed in the wings.

Multipiece slat surfaces are placed on the outer edge of the wing trailing surfaces. The slats consist of ribs attached to the main beam, an aluminium coating and a composite trailing edge. The slats include also the heating pipe used for anti-icing purposes. The slats are reinforced mainly aluminium structures with internal ribs and stiffeners.

Multipart flaps consist of a middle, a front and a rear flap. These three parts are mechanically separated when opening. Each flap part is moved by carriers moving on the flap tracks attached to the bottom of the wing. Flaps usually consist of ribs, spars, honeycomb trailing edge and aluminium plating sheet.

6.3.2.3.3   Spoilers

Spoilers are small, hinged plates located on the top of the wings. Spoilers can be used to slow down an aeroplane (ground spoilers) or to land an aeroplane when deployed on both wings (flight spoiler or speed brakes). Spoilers can also be used to create additional rolling motion.

Ground spoilers are manufactured with a honeycomb structure. Panel surfaces are made in the form of aluminium alloys. Trailing edges are supported by strips. Gasket strips that can be adjusted are placed to the front and rear ends of the spoiler. Flight spoilers are structurally similar to ground spoilers.

6.3.2.3.4   Tabs

The tabs are used to trim or to alter the aerodynamic force on the control surfaces, thereby stabilizing the aircraft in the axis of rotation associated with the main controls. The tabs are hinged to the trailing edges of the main flight control surfaces. The structure of the tab is made in the form of a combination of graphite/epoxy and honeycomb structure.

6.3.3   Empennage

The empennage (generally called tail) is the rear part of the body of the aircraft. The main purpose is to provide stability for the aircraft. Empennage consists of mainly horizontal and vertical stabilizers and tail cone sections.

The tail part of the aircraft is composed of stabilizers and a tail cone. The right and left horizontal stabilizers are also fixed to an adjustable structure from the centre. This movable part can be adjusted by two pivots fixed to the rear pressure chamber. Horizontal stabilizers have to control surface connections on trailing edges. The covering sheet, ribs, spar and centre stabilizer cage frames on the outer side of the horizontal stabilizer form the basic structure.

The vertical stabilizer on the tail part of the plane is a part connected to the body consisting of the covering sheet and ribs connecting the front and back spar. The vertical stabilizer can be removed from the body if required. The leading edge of the stabilizer can also be separated from the basic structure. Front and rear spars, ribs and covering form the main structure of the vertical stabilizer.

The diameter of the tail cone is narrower than the body to maintain the aerodynamic structure of the aircraft. The structure of this section is the same as the body structure of the aircraft. At the back of the tail cone, some APUs are placed on the aircraft. There are vertical and horizontal stabilizer connections on the tail cone. This part was separated from the pressurized area by a bulkhead.

6.3.3.1   Empennage Control Surfaces

6.3.3.1.1   Rudder

The rudder is the main control surface of the flight that controls the rotation around the vertical axis of an aircraft. This action is called "yaw". The rudder is a movable surface mounted on the rear of the vertical stabilizer or fin. It is used to overcome the engine failure in the case of a reversed yaw or a multi-engine aircraft and also allows the aircraft to deliberately slide when necessary. In most aircraft, the rudder is controlled by the pedals of the flight deck. The deviation of the rudder pedal causes the corresponding steering deflection in the same direction. That is, pressing the left steering pedal causes the rudder to deviate to the left.

The rudder consists of front spars, ribs and covering sheets. The rudder hinge connections are located on the front spar. There is a rudder nose behind the vertical stabilizer. The rudder is articulated over ribs located in the rear of the vertical stabilizer.

6.3.3.1.2   Elevators

Elevators are one of the main flight control surfaces that control the movement around the lateral axis of an aircraft. This is called as "pitch" motion. Most aircraft have two elevators which are mounted on the back of each half of the horizontal tail. When manual or autopilot control input is given, the elevators are directed to up or down accordingly.

The lifts react to the forward or reverse movement of the control input. When the pilot pushes the stick forward, the elevator is deflected downwards. This increases the effective chamber of the horizontal tail and increases the lift. Additional lifting on the tail causes rotation around the lateral axis of the aircraft and causes a downward change rotation of the aircraft. The reverse is done by an input of the flight deck controls.

The elevator is basically composed of a double spared on the inside and single spared on the outer side. The entire surface is reinforced with ribs. The elevator hinges are between the horizontal stabilizer connection ribs and the elevator front spar. The elevator is located at the rear end of the horizontal stabilizer structure, with tabs on it.

6.3.4   Landing Gear

The structural elements that the aircraft can stand on for the operation of the landing, departure and taxi movements are called landing gear. The main elements are the caps, pillars, dampers and landing gears.

The landing gears have their own compartments. The rear shock absorber pillars are called main landing gears. The main landing gears carry the main load of the aircraft. The landing gear on the nose was placed on the front section of the aircraft in order to keep the plane in balance and to move in the desired direction.

The absorbers absorb the impulses that occur during landing, take-off and ground motions on the plane. The structure of these absorbers has an outside cylinder and a piston moving therein.

References

1. Vernet J, Firnas AI (1970–1980) Dictionary of scientific biography (CC Gilespie, ed) vol. I, Scribner, New York
2. Wamsley K, Atalay B (2009) Leonardo's universe: the renaissance world of Leonardo da Vinci. National Geographic Society, Washington
3. Niccoli R (2007) History of flight: from the flying machine of Leonardo da Vinci to the conquest of the space. Rizzoli Inc., New York
4. Çelebi E (2003) Seyahatname. Yapı Kredi Kültür Sanat Yayıncılık, Istanbul. (*in Turkish*)
5. Gillispie C (1983) The Montgolfier brothers, and the invention of aviation. Princeton University Press, New Jersey
6. Taylor MJH (1989) Jane's encyclopedia of aviation. Studio Editions, London
7. Anderson JD (2002) The airplane: a history of its technology. American Institute of Aeronautics and Astronautics (AIAA), Reston
8. Pioneers of flight: Otto Lilienthal. [Movie]. Discovery Channel, 8 Jan 2012
9. Dwyer L Sir George Cayley: The Father of Aviation [online]. Available: http://www.aviation-history.com/early/cayley.htm. Accessed 15 Oct 2018
10. Rumerman J (2018) Sir George Cayley - making aviation practical [Online]. Available: http://www.centennialofflight.net/essay/Prehistory/Cayley/PH2.htm. Accessed 15 Oct 2018

11. Gray CF (2018) The five first flights: the slope and winds of big kill devil hill - the first flight reconsidered [Online]. Available: http://www.thewrightbrothers.org/fivefirstflights.html. Accessed 15 Oct 2018
12. Gibbs-Smith CH (1959) Hops and flights: a roll call of early powered take-offs. Flight 75(2619):469
13. Anonymous (1909) Progress of mechanical flight. Flight 1(1):12
14. Baker D (1994) Flight and flying: a chronology, 1st edn. Facts on File, New York
15. Eisenstein P (2004) Biggest jet engine. Pop Mech 181(7):44–46
16. Nahum A (2004) Frank Whittle: invention of the jet. Icon Books, London.; Reprinted edition
17. Dick, R., & Patterson, D. (2010). 50 aircraft that changed the world. Richmond Hill, Ont.: Boston Mills Press
18. Grant R (2017) Flight: the complete history of aviation. DK Publishing, London
19. Machine M (2018) Iconic aircraft of the Vietnam war. Military machine [Online]. Available: https://militarymachine.com/aircraft-vietnam-war/. Accessed 17 Oct 2018

Chapter 7
Technology Review of Sustainable Aircraft Design

Tawfiq Ahmed and Dilek Funda Kurtulus

Since the beginning of the aviation industry, fossil fuel is being used as the only source of fuel to power airplanes. Each year the popularity of aviation industry among the passengers is increasing dramatically because of its short journey time. Therefore, the demand of fossil fuel is also increasing to support the additional need. However, the stock of fossil fuel is reducing and will all be consumed within a couple of decades. In addition to this, the increment of greenhouse gas has become another issue, which should be dealt with urgently in order to not contribute to global warming. Because of these upcoming problems, another source of fuel and a significant development in aircraft design are a must for the future of aviation industry. Some new sources of fuels are being tested as alternative fuels. Solar power, biofuels, and hydrogen fuel are some of them. New techniques for aircraft design have also been developed within the last couple of years. To make these new ideas available, all the aircraft manufacturing companies and engineers should work collectively.

7.1 Introduction

The increasing level of the worldwide emission of greenhouse gases and other pollutants is caused partially by the aviation industry. The severity can be understood by the fact that the growth of the passengers seemed to be doubled by 2050. According to the EU Flightpath 2050, the CO_2 emission level should be reduced by 75% despite the increase in air traffic [1]. To be able to achieve that goal, dramatic improvements on fuel and design efficiency are required along with alternative fuel source.

T. Ahmed · D. F. Kurtulus (✉)
Middle East Technical University, Department of Aerospace Engineering, Ankara, Turkey
e-mail: e194794@metu.edu.tr; funda.kurtulus@ae.metu.edu.tr

© Springer Nature Switzerland AG 2019
T. H. Karakoc et al. (eds.), *Sustainable Aviation*,
https://doi.org/10.1007/978-3-030-14195-0_7

The studies conducted by the Federal Aviation Administration (FAA) show that the USA alone burn 16.2 billion gallons of aviation fuel per year [2]. According to US Environmental Protection Agency, transportation is the second leading sector which contributes 28% of the US greenhouse gas emissions. Only aircrafts account for 9% of all US transportation greenhouse gas emissions [3]. In addition, in Europe the emission of greenhouse gas has increased by 87% from 1990 to 2006 [4]. Therefore, for the future of aviation industry, sustainability and developing green technologies became the most important task.

7.2 General Information

Since the first flight of the Wright brothers, the development in the industry of aviation to date is significant. To cope with the inclining demand from the upgrowing population, the aviation industry has become one of the fastest-growing sectors in the world. Today human beings prefer to fly by an aircraft because it does not only have a short travel time but also a low accident rate than any other transportation system available. The International Air Transport Association (IATA) has published data on how many people fly each year, and within the last 12 years, the number has been doubled (Fig. 7.1) [5].

Therefore, to support this vast need, the number of aircraft produced is also being boosted. Currently, there are almost 500,000 air vehicles including 335,000 active

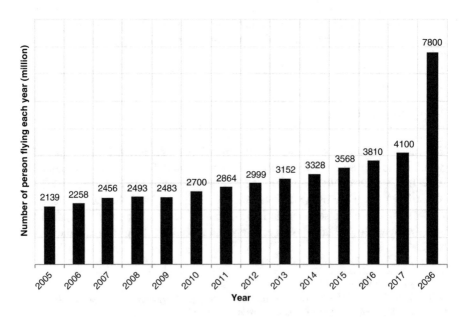

Fig. 7.1 Number of person flying each year since 2005. (Reproduced from Ref. [5])

general aviation aircraft, 18,000 passenger aircraft, 90,000 military aircraft, 27,000 civil helicopters, and 30,000 military helicopters in the world [6].

However, almost all of these aircrafts are fueled by fossil fuel, which is causing some serious damage to our environment by emitting carbon and greenhouse gas. Global aviation causes around 2–3% of the global greenhouse emissions, and the total carbon emission level is crossing the alarming border each day. If aviation industry continues to follow the same way with no strict changes, its threats to global warming are estimated to increase by 10–15% by 2050. In addition to that, noise pollution caused by airplanes became another significant issue due to the expansion of airports and air traffic activities. The adverse reaction from local communities is escalating day by day. On top of that, the reducing storage of underground fossil fuel and its increasing price have become another headache for the airplane building companies. To make the aviation industry more efficient and achieve the goal to reduce the emission by 50% by 2050, the International Civil Aviation Organization has promoted some requirements, and one of them is to design future aircrafts [7].

Given this scenario, to minimize the aviation's emission levels and the aircrafts' noise, aviation industries are seeking for a new design solution which will not only diminish mentioned problems but also will be quite sustainable in terms of fuel consumption. The promising design of new aircraft like Boeing 787 and Airbus A380 can be a doorway to an advanced design technology, which includes usage of new materials for the structure [8]. New developed hybrid materials (like carbon fiber) have been used for lighter aircraft and improved the structure, life-span, and safety requirements at the same time [9]. Less lift and drag production will be faced by an aircraft while using less weight aircraft structure resulting in less thrust needed and less fuel consumption and less carbon emissions. However, weight savings can reduce the emission up to a certain stage even though the aerodynamic improvements are on their best now. Therefore, a sustainable aircraft design, which will use an alternative source of energy other than fossil fuel, is necessary for diminishing the carbon pollution in the environment.

7.3 A Short History of Aircraft Design

History of aviation starts with the concept drawing of Leonardo Da Vinci's flying machine back in the fifteenth to sixteenth centuries [10]. However, the first successful attempt to fly was done by Wright brothers in 1903 in Ohio, USA. That was the world's first practical aircraft and it was named Flyer [11]. Since then until today, significant developments in aircraft design can be observed. In this passage some aircrafts will be mentioned that transformed the aviation history.

In 1909, Blériot XI appeared with a fine technology and established engine in front monoplane tradition with tail-dragger landing gear. This aircraft was built by Louis Blériot and is remembered for crossing the English Channel on July 25 of the same year [12].

In 1911, Deperdussin Monocoque was the first introduced structure with stressed skin in designing aircrafts and became the worldwide standard. This aircraft was developed by Louis Béchéreau and Frederick Koolhoven, and it is significant for launching streamlining revolution which continues until now [13].

In 1913, Sikorsky Ilya Muromets came up with more practical and refined success. This aircraft was developed by Igor Sikorsky, and it was the first multi-engine airplane which had a cockpit with dual controls for two pilots. In addition, this large four-engine biplane had a large cabin with private suites, bed, lavatory, balcony, cabin lighting, and cabin heating [14].

In 1919, Junkers J-13 (F-13) was developed by Hugo Junkers which was known to be the first full metal aircraft with cantilever wing. This aircraft had thick low wings that were able to produce more lift and had an enclosed cabin. This aircraft went for a mass production of around 300 and mostly known as F-13 worldwide [15].

In 1920, Zeppelin-Staaken (Rohrbach) E.4/20 became one of the finest aircrafts which was developed by Adolf Rohrbach. This aircraft had thick stressed skin with a single torsion-box spar for the whole wing making two smooth surfaces: upper and lower. The high-mounted, tapered, cantilever wing had a wingspan of 102 feet. The aircraft had a big cabin to accommodate up to 18 passengers and had enough space for lavatory and luggage. It had a weight of 18,700 pounds and could fly with a cruise speed of 130 mph ranging around 850 miles [16].

In 1930, Douglas DC-1 was developed by Douglas Aircraft Corporation in America and known to be the first scientifically designed aircraft. The twin-engine DC-1 could carry 12 passengers and had a very efficient aerodynamics, controllable pitch propellers, monocoque fuselage, and a retractable landing gear. The aircraft also had a kitchen, plush seats, and restroom for a comfortable flight for the passengers [17].

In 1937, Lockheed XC-35 was the first aircraft in the world that had a pressurized passenger cabin. It could fly with a cabin pressure of 9.5 pounds at an altitude of 33,000 feet [18].

In 1939, Gloster E.28/39 was one the first aircrafts with jet engine. Although it was not the first jet engine-powered aircraft, still its footprint was significant in terms of high-speed transonic or supersonic aerodynamics [19].

In 1947, North American XP-86 Sabre was made to take high-speed advantage of swept-wing structure. The swept wing was built to increase the stability and the speed of the aircraft which was first discovered by a German scientist named Adolf Busemann [20]. In the same year, Bell XS-1 was developed which was the first supersonic aircraft that illustrated the advantages of a thin and low-aspect-ratio wing structure [21].

In 1954, Boeing 367-80 came up with a remarkable jetliner configuration that could fly medium- or long-range flights. The wide-body configuration started from this aircraft which can be widely seen now in Boeing 747 configuration [22].

In 1994, Boeing 777 represented an advanced progress in design, structures, propulsion systems, and flight mechanic technologies. It had twin engines that could produce 90,000 pounds of thrust. The aircraft was designed in computer by using a design program called CATIA (Computer-Aided Three-Dimensional Interactive

Application). It was built to compete with two other rivalry aircrafts named the Airbus A330/340 and the McDonnell Douglas D-11 [23].

All of these aircrafts had shown the future of modern aircraft design and inspired designers and engineers to come up with new design solutions in order to get more efficient performances.

7.4 Sustainable Aircraft Design in Literature

In literature, sustainable aircraft design can be categorized depending on the fuel used by those aircrafts. In this chapter four types of aircraft designs have been discussed to give a better idea about the design concepts, namely, solar-powered aircrafts, hydrogen-fueled aircrafts, biofueled aircrafts, and hybrid aircrafts.

7.4.1 Solar-Powered Aircrafts

The recent developments in the areas of energy storage and solar cells have opened a new door to solar-powered aircraft design. Even though these aircrafts are mostly designed for unmanned applications and can carry relatively small payloads, they are very useful in some areas such as communications, atmosphere monitoring, and reconnaissance [24].

The wings of the aircraft are covered with solar panels which retrieve the solar energy from the sunlight during the daytime to supply energy to the propulsion system, the control electronics, and the rechargeable battery (Fig. 7.2). During the night, the rechargeable battery becomes the only source of energy which discharges slowly until morning when a new cycle starts again [25].

Although the concept is pretty straightforward, due to the boundaries of technology, it has been a tough path to build an aircraft which can carry human beings. At first it was all about making an unmanned aircraft of which the only purpose was to

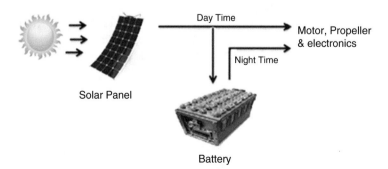

Fig. 7.2 Solar power distribution in aircraft

fly for seconds. By course of time, advanced technology altered the seconds to minutes, minutes to hours, and hours to days. Now the main purpose of solar-powered aircraft is not only to fly for days but also to carry human beings along with it. The progress in this field opened a new door to think of solar power as an alternative source of fossil fuel.

In recent years many initiatives have been taken to replace fossil fuel with solar power. For instance, Solar Impulse is a Swiss privately funded project of which the main objective is to design and build long-range solar-powered aircraft that can carry human beings. During the project, two fixed-wing aircrafts, namely, Solar Impulse 1 and Solar Impulse 2, are built [26]. The first aircraft, Solar Impulse 1, was the first prototype that could stay airborne up to 36 hours [27]. Its first test flight was in December 2009. In July 2010, the aircraft flew for 26 hours (including 9 hours night flight) from Switzerland to Spain [28]. The second aircraft, Solar Impulse 2, was bigger and carried more solar cells and more powerful motors than the previous one. Both of these aircrafts were piloted by Piccard and Borschberg only. Starting in March 2015, within 16 months Solar Impulse 2 was able to make the first circumnavigate of the earth completing approximately 42,000 km using only solar power [29]. The aircrafts are single-seated monoplanes of which the source of power is photovoltaic cells. The concept of designs for the both aircrafts is same which has focused on long wingspan to contain enough space for necessary solar cells and light carbon-based slender body to sustain the power produced from the solar cells. For instance, the wingspan of Solar Impulse 2 is 71.9 m, which is a little bit shorter than the world's largest passenger aircraft Airbus A380 [30] but weighs only around 2.3 tons, while A380 weighs 500 tons [31].

Another example of manned solar-powered aircraft is Sunseeker Duo. This aircraft is capable of flying with two people onboard. The structure of this aircraft is ultralight and aerodynamically very efficient which enables it to fly using only solar power. The battery pack is located in the fuselage to store energy extracted from the solar panels inserted on its wings and tail surfaces. The wings of this aircraft can be fully folded and even packed into a trailer to be able to accommodate into a hanger [32, 33]. The first flight was conducted in December 2013. So far, it has logged several hundred hours in the air with more than a few passengers.

The first cross-country flight is performed in 2014. In 2015, the Alps were crossed with this aircraft using solar power only. This flight across the Alps set up a milestone in terms of solar-powered aircraft and renewable energy sources [32].

Table 7.1 shows the comparison of the design specifications and performance characteristics of the example solar-powered manned aircrafts underlined.

7.4.2 Hydrogen-Fueled Aircrafts

Hydrogen fueled aircrafts are now preferred as sustainable aircrafts because they produce no emissions other than water, and thus they are considered to be environment friendly. Commercial aircrafts are now being tried to be powered by hydrogen

Table 7.1 Characteristics comparison of solar-powered aircrafts in literature from Refs. [32, 34–36]

		Solar Impulse 1	Solar Impulse 2	Sunseeker Duo
General characteristics	Crews + capacity	1	1	2
	Length (m)	21.85	22.4	N/A
	Wingspan (m)	63.4	71.9	22
	Height (m)	6.40	6.37	N/A
	Wing area (m²)	200	269.5	N/A
	Empty weight (kg)	N/A	N/A	290
	Maximum weight (kg)	2000	2300	470
	Takeoff speed (km/h)	35	36	N/A
	Number of motors	4	4	1
	Propeller diameter(m)	3.5	4	N/A
Performance	Cruise speed (km/h)	70	90 60 at night	72
	Maximum speed (km/h)	N/A	140	161 $(V_{ne})^a$
	Service ceiling (m)	8500 12,000 (max)	8500 12,000 (max)	3810 3962 (max)

$^aV_{ne}$ not exceed velocity

fuel cells or liquid hydrogen. To prevent greenhouse gas emission in the environment, researches are being continued to achieve more development in this sector.

Hydrogen fuel can be used in two different ways as hydrogen fuel cell or liquid hydrogen fuel.

7.4.2.1 Hydrogen Fuel Cell

Using hydrogen fuel cell in aircraft has been a new concept for sustainable aircraft design. Recently, Boeing has built a propulsion system which is powered by fuel cell to fly manned aircraft which was converted from a glider into a fuel cell aircraft [37]. The aircraft's electric propeller was driven by two power sources, a PEM (proton-exchange membrane) fuel cell and a Li-ion (lithium-ion) battery. The fuel cell supplies the main power, and the batteries supply the extra power needed during the takeoff and the climbs [37].

The concept of the hydrogen fuel cell is pretty straightforward: the fuel cell system produces electrical energy from chemical energy which is extracted from the reaction between hydrogen and oxygen, and then to reduce the weight of the system, the produced hydrogen gas is stored under high pressure, and the needed oxygen is collected from the environment [37].

Li-ion batteries are the auxiliary power source of the aircraft. The duty of these batteries is to supply the needed additional power during takeoff and climb. When the aircraft reaches its cruise altitude, the power line is disconnected to save the power. During the level flight, the whole power comes only from the hydrogen fuel cell. However, the Li-ion batteries are rechargeable, and they have to be fully charged before each flight to ensure an undisturbed flight [37].

7.4.2.2 Liquid Hydrogen (LH₂)

Another way of using hydrogen as fuel is highly pressurized liquid hydrogen (LH_2). The reason behind considering LH_2 as a future fuel is that it has higher gravimetric energy density than conventional kerosene (33.3 kWh/kg and 12 kWh/kg, respectively). Also in terms of no carbon and greenhouse emission, LH_2 can be handful as a renewable energy source. In between liquid hydrogen (LH_2) and gaseous hydrogen ($G H_2$), LH_2 is more feasible because it offers a specific volume of 5.6 times less than $G H_2$ [38].

Talking about conventional jet fuel, Jet-A can be an example. In the USA Jet-A fuel is mostly being used [39], and Table 7.2 shows the comparison between the liquid hydrogen and Jet-A fuel in terms of some selected thermodynamic parameters.

The design of the aircraft using liquid hydrogen fuel has been named as Cryoplane [42]. A lot of researches are being conducted to optimize the feasible design of a Cryoplane in order to fit it in the current aviation industry although the huge volume of the LH_2 makes it harder for the engineers to build an aircraft which is capable of carrying such huge volume without losing aerodynamic and structural characteristics.

7.4.3 Biofueled Aircrafts

In the aviation industry, biofuels are also known as bio-jet fuels. In order to achieve the goal to reduce the carbon emission by 2050, they are assumed to be the only option. Natural sustainable sources like animal fats and vegetable oils can be used to produce bio-jet fuels. The most promising news is that to be able to use these fuels, the existing jet engines do not need any kind of modifications [43].

To support this attempt to reduce the carbon emission, NASA predicts that using 50% aviation biofuel mixture can reduce the carbon pollution by 50–70% [44]. In July 2011, biofuels were approved for commercial use after doing some researches by aircraft manufacturers, engine makers, and oil companies. Some commercial test flights have been conducted by top airliners using biofuels to support the movement [45].

Since solar-, electric-, and hydrogen-powered aircrafts are not expected to be feasible within a very short time, biofuel has become the light in the dark tunnel because of its high power-to-weight ratio. To follow this path, the International Air Transport Association (IATA) has been supporting all kinds of research and development of alternative fuels [46]. In addition, in 2008, a group is formed by interested

Table 7.2 Comparison of properties of Jet-A fuel [40] and liquid hydrogen [41] at normal boiling point

Parameter	Jet-A ($CH_{1.93}$)	LH_2
Temperature (K)	439.817	20.369
Density (kg/m³)	810.53	70.9
Molecular weight (kg/mol)	168×10^3	2.016×10^3
Heat combustion (J/kg)	42.798×10^6	120×10^6
Specific heat capacity (J/kg K)	1.968×10^3	9.747×10^3

Table 7.3 Flights using biofuels

Operator	Aircraft model	Biofuel used	Date
SpiceJet Airlines, India [48]	Bombardier Q400	Nonedible oils, agricultural wastes, and biodegradable residues of industrial wastes	August 28, 2018
Singapore Airlines [49]	Airbus A350-900	Hydro-processed esters and fatty acids (HEFA)	May 1, 2017
KLM [50]	Boeing and Embraer	Waste vegetable oils	September 8 and March 31, 2016
Hainan Airlines [51]	Boeing 737-800	Waste vegetable oils	March 21, 2015
Scandinavian Airlines [52]	Boeing 737-700	Waste vegetable oils	November 11, 2014

airlines named Sustainable Aviation Fuel Users Group (SAFUG) to support the idea of using biofuels more widely. Several NGOs have also given their hands for establishing this group such as the Natural Resources Defense Council and Roundtable on Sustainable Biofuels (RSB). The member airlines contribute more than 15% of the aviation industry, and all the CEOs of these member airlines have come together to develop and deploy sustainable biofuels for aviation industry [46, 47].

During the recent years, to support the idea of using biofuels, some airlines have stepped up to use biofuels for both commercial and test flights. Table 7.3 lists recent flights with biofuels.

7.4.4 Hybrid Aircraft Technologies

7.4.4.1 Laminar Flow Control

The brightest technology to reduce aerodynamic drag is laminar flow control (LFC). This technique reduces drag using boundary layer suction on wetted surfaces such as the empennage, the engine nacelles, and the fuselage of the future aircrafts [53]. Resizing the aircraft and redesigning the wings can cause a total aerodynamics drag reduction by 50% more than the current aircraft designs.

Reduction of total drag means reduction of power requirements. Therefore, use of LFC method can open up some new doors to electric propulsion systems like hydrogen fuel cells and batteries [54].

7.4.4.2 Blended Wing Body (BWB) Configuration

The blended wing body configuration is a way of designing in order to make aircrafts aerodynamically more efficient. Aerodynamically more efficient means less use of fossil fuel which leads to an environmentally friendly aviation.

The way of designing an unconventional unique configuration follows the same way of designing traditional approach and also novel method. Basically Breguet's range equation (Eq. 1) is used to design the preliminary sizing which is:

$$\text{Range} = \frac{V}{g} * \frac{1}{\text{SFC}} * \frac{L}{D} \ln \frac{w_{\text{initial}}}{w_{\text{final}}} \tag{1}$$

where SFC is specific fuel consumption, L/D is lift-to-drag ratio, V is velocity, w_{initial} is initial weight, and w_{final} is final weight.

The most effective part of the equation is the lift-to-drag ratio. The more the lift-to-drag ratio is, the less the aircraft needs power to fly. In BWB aircrafts the lift-to-drag ratio is pretty high which makes these aircrafts fuel efficient, and it decreases the fuel burn per passenger per km significantly. In addition, BWB aircraft's unibody design enables it to have lower empty weight than that of conventional aircrafts [55].

Recently, NASA and Boeing are developing a prototype of BWB aircraft named X-48 in NASA Langley Research Center in Virginia [56].

Lockheed Martin, another aircraft manufacturing company, has also developed some high-altitude and long-endurance (HALE) UAVs using blended wing **body** concept named RQ-3 DarkStar and RQ-170 Sentinel. The RQ-170 Sentinel is developed by Skunk Works, which works for Lockheed Martin Corporation [57].

China is not so behind in this design technique as they have also developed a UAV named AVIC 601-S. The UAV is mainly designed for low observation by Shenyang Aircraft Design Institute (SYADI) and Shenyang Aerospace University [58].

7.4.4.3 Gull-Boxed Wing

The main purposes of gull and boxed design are to improve the aircraft's structural support, block aeroelastic effects, and reduce induced drag. These types of unique aircraft designs reduce energy consumption and increase aerodynamic efficiency because of two major reasons.

Firstly, the relatively tall winglets connect the main and aft wing which helps to minimize wingtip vortices formation enabling the aircraft to be induced drag-free.

Secondly, the boxed wing of these aircrafts creates a firm structure of the whole aircraft and significantly decreases aeroelastic effects like divergence and flutter [59]. In addition, more wing volume is produced due to the connecting wings which allow more lift during flights and enable to carry more fuel tanks.

One of the experimental aircrafts built is Pyxis which is showing the door to the future aviation because of its advanced aerodynamics, its strong structure and promising cabin interiors, and its hydrogen fuel-based propulsion system. The design of Pyxis was conducted by the Mechanical and Aerospace Engineering Department at the University of California, Davis (USA) [60]. The goal set by NASA to reduce energy consumption up to 80% by 2045 is met by the design of Pyxis.

Table 7.4 Boeing 737-900/Pyxis comparison [62]

Characteristics	Boeing 737-900	Pyxis
Maximum takeoff weight (kg)	74,390	52,616
Range (km)	5500	6450
Passengers	189	189
Cargo volume (m³)	52	52
Wingspan (m)	35.8	37.2
Cruise thrust-specific fuel consumption (TSFC) (kg/Nh)	0.06447	0.02293

To achieve the goal of reducing energy consumption, the aircraft's design includes more advanced and efficient box wing and environmentally friendly liquid hydrogen (LH_2) fuel.

This aircraft design is a perfect platform to use more LH_2 that will replace the use of fossil fuels like kerosene. LH_2 has 2.78 times the amount of energy content per unit mass compared to conventional fossil fuels. However, to produce the same energy as the fossil fuels, LH_2 requires four times the volume [61]. The design will be able to reduce NOx emissions by 60–90% in the future compared to the present time.

The Pyxis can be compared with Boeing 737-900 as shown in Table 7.4. It is observed that the energy consumption of Pyxis is 80% lower than the conventional aircraft.

7.5 Aircraft Design Methodology

Aircraft design methodology is the way an aircraft is designed by engineering design processes to meet various design requirements such as manufacturer demands, structural and economic boundaries, and safety requirements. The methodology can be divided into three phases: conceptual design, preliminary design, and detailed design.

7.5.1 Conceptual Design Phase

Conceptual design phase includes sketching different aircraft configurations which meet the design requirements and specifications. In this phase, the main goal of a designer is to get the optimized design that meets all the specified requirements in terms of aerodynamics, structures, propulsion system, and flight performance [63]. Important design fundamental aspects such as wing configuration, fuselage size and shape, engine selection, and tail configuration are decided in this phase. At the end of this phase, a layout of the whole aircraft configuration is designed on paper or computer to be analyzed later on by other engineers and designers.

7.5.2 Preliminary Design Phase

Preliminary design phase involves the further analysis of the conceptual design configuration. Conceptual design is refined in preliminary design stage. Testing is initiated on aerodynamics (e.g., wind tunnels, CFD analysis), propulsion, structure, and stability control [63]. Structural and control systems are analyzed and corrected in this phase in case of any instabilities to get the final design configuration. After finalizing the design configuration, manufacturer decides whether to proceed with the production of the aircraft [63]. All the aerodynamic, propulsion, structural, control, performance, and safety aspects are covered in this phase.

7.5.3 Detailed Design Phase

The detailed drawings of all the components of the aircraft are released in this phase [64]. Detailed design phase handles the manufacturing of the aircraft. This phase also includes the certification process and all component and system tests. Moreover, flight simulators are developed for the aircraft at this phase.

7.5.4 Aspects of Sustainable Aircraft Design

The main aspects of aircraft design are,

(a) Aerodynamics
(b) Propulsion
(c) Weight and structure
(d) Control

To achieve the design requirements, these factors play the major roles [65].

Aerodynamics of an aircraft mostly depends on the configuration of wings and fuselage. Almost all of the lift is produced by wings. In three ways the wings can be mounted to the fuselage: high, middle, and low. Some important parameters of wings are aspect ratio, taper ratio, thickness to chord ratio, sweep angle and dihedral angle [66]. The cross-sectional shape is called airfoil, and its profile can change according to the mission requirement.

The fuselage of an aircraft includes the cockpit, passenger cabin, and cargo hold [63].

Propulsion of an aircraft can be obtained by specially designed aircraft engines or electric motors. The main parameters of aircraft engine are maximum engine thrust, fuel consumption, engine geometry, and engine mass.

Weight and structure of an aircraft are interrelated. Structure of an aircraft should be designed to withstand any kind of stresses during flight such as cabin pressurization, turbulence, and engine or rotor vibration. The aircraft structure focuses on

Table 7.5 Design comparison between sustainable aircrafts

		Solar powered	LH$_2$ fueled	Biofuel powered	BWB aircrafts
Aerodynamics	Wing	Large wing span (to accommodate solar panels)	No specific change needed	No specific change needed	Wing and body are blended together
	Fuselage	Fuselage with flat surface to easily mount the cells	Large fuselage is needed to accommodate fuel tanks	No specific change needed	
Propulsion		Electric motors powered by batteries that get energy from solar cells	Modified turboprop and turboshaft engines	Conventional turboprop, turboshaft, or jet engines	Electric motors
Weight and structure		Carbon fiber-based and ultralight structure	Strong and huge to support fuel tanks	No specific change needed	Carbon fiber-based structure

strength, stiffness, fatigue (durability), stability, toughness, fail safety, corrosion resistance, maintainability, and ease of manufacturing [67].

To design these components for a sustainable aircraft, some facts should be considered. Table 7.5 shows the comparison between solar-powered, liquid hydrogen (LH$_2$)-powered, biofuel-powered, and blended wing body (BWB) aircrafts.

From Table 7.5, it can be seen that biofueled aircrafts do not need any kind of modifications to be able to use biofuels [43] and that can save a lot of money. Solar-powered aircraft needs a very large wingspan and a very strong and lightweight structure which creates a challenge for the designers and engineers. Liquid hydrogen-fueled aircraft needs a huge fuselage to store all its LH$_2$ tanks, and the engines should also be modified to be able to use LH$_2$, and these modifications can increase the cost and effort. Blended wing body configuration on the other side is normally used for small unmanned aerial vehicles.

7.6 Conclusion

Today the global aviation industry is far more concerned about the environment than before and is willing to keep on funding for new researches and developments. The increasing level of carbon and greenhouse gas emission needs to be stopped or reduced. To fight this situation, new advancements are taking place such as new aircraft design concepts using sustainable energies like solar power, hydrogen fuel cell, liquid hydrogen, and biofuel in order to replace fossil fuel. In addition, hybrid aircrafts are being designed and tested to increase the efficiency of aircrafts with laminar flow control technology, blended wing body configuration, and gull-boxed wing configuration. In other words, all these technologies should be employed collectively to get the maximum result possible in designing sustainable aircrafts.

References

1. European Commission (2011) Flightpath 2050: Europe's vision for aviation. Retrieved from: https://ec.europa.eu/transport/sites/transport/files/modes/air/doc/flightpath2050.pdf. Accessed 1 Oct 2018
2. Federal Aviation Administration, FAA Aerospace Forecast (2016) Total jet fuel and aviation gasoline fuel consumption. p. 76
3. U.S. Environmental Protection Agency (2016) Fast facts on transportation greenhouse gas emissions. Retrieved from https://www.epa.gov/greenvehicles/fast-facts-transportation-greenhouse-gas-emissions. Accessed 1 Oct 2018
4. European Commission (2006) Climate change: commission proposes bringing air transport into EU emissions trading scheme (press release). EU press release. Retrieved from: http://europa.eu/rapid/press-release_IP-06-1862_en.htm. Accessed 1 Oct 2018
5. The International Air Transport Association (IATA) (2017) 2036 Forecast reveals air passengers will nearly double to 7.8 billion. Retrieved from https://www.iata.org/pressroom/pr/Pages/2017-10-24-01.aspx. Accessed 1 Oct 2018
6. Agarwal R (2018) Sustainable (green) aviation: challenges & opportunities. International symposium on sustainable aviation, Rome, 9–11 July, 2018, p. 4
7. Maurice L, Lee D (2009) Assessing current scientific knowledge, uncertainties and gaps in quantifying climate change, noise and air quality aviation impacts, final report of International Civil Aviation Organization (ICAO) Committee on Aviation and Environmental Protection (CAEP) workshop. US Federal Aviation Administration and Manchester Metropolitan University, Washington DC/Manchester
8. Cikovic A, Damarodis T (2012) The Boeing 787's role in new sustainability in the commercial aircraft industry. University of Pittsburgh. Retrieved from: https://www.researchgate.net/publication/272026500_The_Boeing_787%27S_Role_in_New_Sustainability_in_the_Commercial_Aircraft_Industry. Accessed 1 Oct 2018
9. Immarigeon J, Holt R, Koul A, Zhao L, Wallace W, Beddoes J (1995) Lightweight materials for aircraft applications. Mater Charact 35(1):41–67
10. Pioneer profile. Leonardo Da Vinci (1452–1519). Retrieved from: https://www.aiaa.org/secondarytwocolumn.aspx?id=15129. Accessed 1 Oct 2018
11. Tise LE (2009) Conquering the sky. Palgrave MacMillan, New York, p 22
12. Gollin A (1989) The impact of air power on the British people and their government. Stanford Unversity Press, Stanford, California, pp 70–74
13. Mirguet H (1912) Le Monocoque Deperdussin. L'Aérophile XX(28):410–411
14. Sikorsky II (1967) The story of the Winged-S: an autobiography. Dodd, Mead & Company, New York, pp 69–117
15. Dienel HL, Schiefelbusch M (2000) German commercial air transport until 1945. Revue belge de philologie et d'histoire 78(3–4):955–956
16. Spooner S (1921) The Zeppelin-Stakken all-metal monoplane. pp 185–186. Retrieved from: https://www.flightglobal.com/pdfarchive/view/1921/1921%20-%200185.html
17. DC-1 commercial transport. Historical snapshot. Retrieved from: https://www.boeing.com/history/products/dc-1.page. Accessed 1 Oct 2018
18. Robinson D (1973) The dangerous sky. University of Washington Press, Seattle, pp 103–104
19. Nahum A, Whittle F (2004) Invention of the jet, Chapter 3. Icon Books, London
20. Hallion RP (1979) Lippisch, Gluhareff, and Jones: the emergence of the delta planform and the origins of the sweptwing in the United States. Aerospace Historian 26(1):1–10
21. U.S. Air Force (1948) Air force supersonic research airplane XS-1 report no. 1, Wright-Patterson Air Force Base, pp 22–26
22. Linden FRVD, Spencer AM, Paone TJ (2016) Milestones of flight: the epic of aviation with the National Air and Space Museum. National Air and Space Museum, in association with Zenith Press, Washington, DC, pp 148–153
23. Modern Airliners (2018) Boeing 777 history. Boeing triple seven history. Retrieved from: http://www.modernairliners.com/boeing-777/boeing-777-history

24. Phillips WH (1980) Some design considerations for solar-powered aircraft, NASA technical paper 1675
25. Noth A, Siegwart R (2006) Design of solar powered airplanes for continuous flight. ETH, Zürich, p 1
26. Div S (2016) Solar impulse 2: the groundbreaking aircraft demonstrating the possibilities of clean energy, The Independent
27. HB-SIA Mission (2011) Solar impulse project. Retrieved from http://www.solarimpulse.com/en/documents/hbsia_mission.php?lang=en&group=hbsia. Accessed 1 Oct 2018
28. Malcolm C (2010) Swiss solar plane makes history with night flight. Retrieved from http://www.swisster.ch/news/science-tech/swiss-solar-plane-makes-history-with-night-flight.html. Accessed 1 Oct 2018
29. Batrawy A (2015) Solar-powered plane takes off for flight around the world. Retrieved from https://www.msn.com/en-us/news/technology/solar-powered-plane-takes-off-for-flight-around-the-world/ar-AA9wVrL. Accessed 1 Oct 2018
30. Diaz J (2007) Solar impulse: around the world in a 100% sun-powered airplane. Retrieved from https://gizmodo.com/262940/solar-impulse-around-the-world-in-a-100-sun-powered-airplane. Accessed 1 Oct 2018
31. Al-Jazeera (2015) Solar-powered Swiss plane attempts flight around world. Retrieved from https://www.aljazeera.com/news/2015/03/solar-impulse-swiss-plane-uae-150309032941002.html. Accessed 1 Oct 2018
32. Sunseeker Duo – first two seat solar powered aircraft. Retrieved from: https://www.solar-flight.com/sunseeker-duo. Accessed 1 Oct 2018
33. Sun powers first two-place electric aircraft. Sport Aviation. 14 July 2014
34. Diaz J (2007) Solar impulse: around the world in a 100% sun powered airplane. Retrieved from: https://gizmodo.com/262940/solar-impulse-around-the-world-in-a-100-sun-powered-airplane. Accessed 1 Oct 2018
35. Solar Impulse (2014) Building a solar aircraft. Retrieved from https://solarimpulse.com. Accessed 1 Oct 2018
36. Sigler D (2014) Sunseeker duo goes dual. Retrieved from: http://sustainableskies.org/sun-seeker-duo-goes-dual. Accessed 1 Oct 2018
37. Lapena-Rey N, Mosquera J, Bataller E (2008) Environmentally friendly power sources for aerospace applications. J Power Sources 181:353–362
38. Mital SK, Gyekenyesi JZ, Arnold SM, Sullivan RM, Manderscheid JM, Murthy PLN (2006) Review of current state of the art and key design issues with potential solutions for liquid hydrogen cryogenic storage tank structures for aircraft applications. National Aeronautics and Space Administration, Cleveland
39. Jet A/Jet A-1. Retrieved from: https://www.exxonmobil.com/en/aviation/products-and-services/products/jet-a-jet-a-1. Accessed 1 Oct 2018
40. NIST (2017) Thermophysical properties of fluid systems. National Institute of Standards and Technology, Gaithersburg
41. Sehra AK, Whitlow W Jr (2004) Propulsion and power for 21st century aviation. Prog Aerosp Sci 40:199–235
42. Cryoplane (2002) Liquid hydrogen fuelled aircraft- system analysis (Cryoplane). Retrieved from: https://cordis.europa.eu/project/rcn/52464_en.html. Accessed 1 Oct 2018
43. Yakovlieva A, Vovk O, Biochenko S, Lejda K (2018) Evaluation of jet engine parameters using conventional and alternative jet fuels. International symposium on sustainable aviation, 9–11 July 2018, Rome, p. 31
44. Megan E (2017) NASA confirms biofuels reduce jet emissions. Retrieved from: https://www.flyingmag.com/nasa-confirms-biofuels-reduce-jet-emissions. Accessed 1 Oct 2018
45. Louise D (2011) Airlines win approval to use biofuels for commercial flights. Retrieved from: https://www.bloomberg.com/news/articles/2011-07-01/airlines-win-approval-to-use-plant-based-biofuels-on-commercial-flights. Accessed 1 Oct 2018
46. Alternative jet fuel. International Air Transport Association (IATA), alternative fuel strategic partnerships. Retrieved from: https://www.iata.org/whatwedo/environment/Pages/sustainable-alternative-jet-fuels.aspx. Accessed 1 Oct 2018

47. Sustainable Aviation Fuel Users Group (2012). Retrieved from: http://www.safug.org/ information/pledge. Accessed 1 Oct 2018
48. Kavita U (2018) SpiceJet flies India's first biofuel flight from Dehradun to Delhi. Indian Express. Retrieved from: https://indianexpress.com/article/business/aviation/spicejet-oper- ates-indias-first-biofuel-powered-flight-5326913. Accessed 1 Oct 2018
49. Singapore Airlines (2017) SIA and CAAS partner to operate first 'green package' flights in the world. Retrieved from: https://www.singaporeair.com/en_UK/es/media-centre/press-release/ article/?q=en_UK/2017/April-June/jr1117-170503. Accessed 1 Oct 2018
50. KLM (2016) KLM to operate biofuel flights out of Los Angeles. Retrieved from: https://news. klm.com/klm-to-operate-biofuel-flights-out-of-los-angeles. Accessed 1 Oct 2018
51. Boeing, Hainan Airlines operate China's first cooking oil-powered flight. Retrieved from: https://cleantechnica.com/2015/03/25/boeing-hainan-airlines-operate-chinas-first-cooking- oil-powered-flight. Accessed 1 Oct 2018
52. Scandinavian Airlines (2014) SAS tar av med biofuel. Retrieved from: https://www.sasgroup. net/en/sas-tar-av-med-biofuel/
53. Beck N., Landa T., Seitz A., Boemans L., Liz Y., Radespiel R.: Drag reduction by laminar flow control. In: Energies 11(1), S252. https://doi.org/10.3390/wn11010252. (2018)
54. Kadyk T, Winnerfeld C, Hanie RR, Krewer U (2018) Analysis and Design of Fuel Cell Systems for aviation. Energies 11(2):B375. https://doi.org/10.3390/en11010166
55. Patel N (2018) Can configuration alone make for a greener aircraft? The case for a blended wing-body, medium size, medium range transport. International symposium on sustainable aviation, 9–11 July 2018, Rome, p. 27
56. NASA & Boeing (2008) The X-48B blended wing body. Available at: https://www.nasa.gov/ vision/earth/improvingflight/x48b.html. Accessed 1 Oct 2018
57. RQ-170 sentinel unmanned aerial vehicle. Retrieved from: https://www.airforce-technology. com/projects/rq-170-sentinel. Accessed 1 Oct 2018
58. General R (2016) China unveils their impressive prize-winning attack drone. Retrieved from: https://nextshark.com/china-unveils-impressive-prize-winning-attack-drone-ever. Accessed 1 Oct 2018
59. Khalid A, Kumar P (2014) Aerodynamic optimization of box wing – a case study. International Journal of Aviation, Aeronautics, and Aerospace 1(4):1–45
60. Pnithan J, Brenna J, Janine M, Shreya R, Khaschuchuluun T (2018) PYXIS: ultra-efficient commercial aircraft with gull-boxed wings and liquid hydrogen fuel. International symposium on sustainable aviation, 9–11 July 2018, Rome, p. 73
61. Erikson P Primary fuels for energy conversion. MAE 218. Davis, CA. Lecture 4
62. Roskam J (1989) Airplane design parts I–VIII. Roskam Aviation and Engineering Corporation, Ottawa
63. Raymer D (1992) Aircraft design - a conceptual approach. American Institute of Aeronautics and Astronautics, Washington, DC. isbn:0-930403-51-7
64. Anderson JD (1999) Aircraft performance and design. McGraw Hill, Singapore. isbn:0-07-001971-1
65. Kroo I (2001) Techniques for aircraft configuration optimization. In: Aircraft design: synthesis and analysis. Stanford University, Stanford
66. Lloyd RJ, Rhodes D, Simpkin P (1999) Civil Jet Aircraft Design. Arnold Publishers, London, UK
67. Megson THG (2007) Aircraft structures for engineering students, Elsevier aerospace engineer- ing series, 4th edn. Butterworth-Heinemann, Oxford/Burlington. isbn:978-1-85617-932-4

Index

© Springer Nature Switzerland AG 2019
T. H. Karakoc et al. (eds.), *Sustainable Aviation*,
https://doi.org/10.1007/978-3-030-14195-0

Printed in the United States
By Bookmasters